建筑的着陆艺术
The Art of Landing

[荷兰] MVRDV建筑设计事务所等 | 编

李璐 | 译

大连理工大学出版社

004	从超现代主义到元现代主义 _ Hans Ibelings

世界前沿的阿拉伯建筑

008	世界前沿的阿拉伯建筑 _ Diego Terna
016	珍珠场地，博物馆及珍珠之路入口 _ Valerio Olgiati
030	迪拜购物中心苹果店 _ Foster + Partners
040	卡塔尔国家博物馆 _ Ateliers Jean Nouvel
062	卡塔尔国家图书馆 _ OMA

建筑的着陆艺术

084	建筑的着陆艺术 _ Nelson Mota
090	奥乔克布拉塔斯住宅 _ ELEMENTAL
110	Bosc d'en Pep Ferrer住宅 _ Marià Castelló
130	Casa Biblioteca住宅 _ Atelier Branco Arquitetura
146	菲尔德霍夫住宅 _ Pavol Mikolajcak Architekt

(不)熟悉的地方

160	(不)熟悉的地方——在非常规的经验中寻找意义 _ Angelos Psilopulous
166	印记 _ MVRDV
180	UCCA沙丘艺术博物馆 _ OPEN Architecture
194	Tij观测站 _ RAU Architects + RO&AD Architecten
208	AZULIK Uh May艺术中心和IK实验室 _ Roth (Eduardo Neira)

220	建筑师索引

004 From Supermodernism to Metamodernism_Hans Ibelings

Arab Architecture in front

008 Evolution on the City: Arab Architecture in front of the World_Diego Terna

016 Pearling Site, Museum and Entrance to the Pearling Path_Valerio Olgiati

030 Apple Dubai Mall_Foster + Partners

040 National Museum of Qatar_Ateliers Jean Nouvel

062 Qatar National Library_OMA

Houses, the Art of Landing

084 Houses, the Art of Landing_Nelson Mota

090 OchoQuebradas House_ELEMENTAL

110 Bosc d'en Pep Ferrer_Marià Castelló

130 Casa Biblioteca_Atelier Branco Arquitetura

146 Felderhof House_Pavol Mikolajcak Architekt

(Un)familiar Places

160 (Un)familiar Places – Finding Meaning in Non-Routine Experiences_Angelos Psilopulous

166 The Imprint_MVRDV

180 UCCA Dune Art Museum_OPEN Architecture

194 Tij Observatory_RAU Architects + RO&AD Architecten

208 AZULIK Uh May and IK LAB_Roth (Eduardo Neira)

220 Index

从超现代主义到元现代主义
From Supermodernism to Metamodernism

Hans Ibelings

　　《超现代主义:全球化时代的建筑》一书问世已有二十多年。时至今日,这本书仍然是我唯一一本畅销书。每当有人让我说点什么的时候,我就觉得自己像一个上了年纪的摇滚歌手,即使观众只对这首很久以前的热门歌曲感兴趣,我仍然会推出新歌。

　　即使我不愿意重新审视自己的作品,我在《超现代主义》的介绍中还是写了一些自认为仍然合理有效的东西,那就是,对当代建筑的每一次解读都是暂时的,用美国实用主义哲学家查尔斯·皮尔斯的话来说,每一次解读都是最好的猜测。

　　让最好的猜测保持20年都是合理的,这太长了。即使我先前所写的关于超现代主义的文章仍然有一定的道理,但回过头来看,我当时的想法显然还不够成熟,我不该忽视后现代主义。

　　对于超现代主义的基本说法是,它是在后现代主义及其衰落的分支解构主义之后,出现的一种与后现代主义截然不同的建筑类型。超现代主义不关心周围的环境,而是关注情境;它优先考虑的是体验,而不是意义;它提倡一种中立的抽象,而不是具有象征意义的形式和姿态。简而言之,如果博物馆是最优秀的后现代主义建筑代表,那么机场就是与之相当的超现代主义建筑代表。

　　1990年以来,OMA、赫尔佐格&德梅隆、伊东丰雄、让·努维尔等人带有后现代主义倾向的抽象作品都被我称作了超现代主义,该词是借用了马克·奥格一本优质小书《非之地:超现代性导论》中的概念。

　　我当时宣称,超现代主义已经取代了后现代主义,这一说法其实并不成熟,现在看来,超现代主义或许只是后现代

It is now more than twenty years ago that Supermodernism: Architecture in the Age of Globalization came out. It is still my only book that became a hit of sorts. Whenever people ask me to say something about it, I feel like an aging rocker, who still brings out new songs, even if the audience is only interested in this one hit from long ago. Even if I am reluctant to revisit my own work, I wrote in the introduction of Supermodernism something which I consider to be still valid, namely that every reading of contemporary architecture is provisional, a best guess, in the terminology of the American pragmatist philosopher Charles Peirce.

Twenty years is an eternity for best guesses to keep their validity, and even if I think there is still a certain truth in what I wrote about supermodern, even if it is evident in retrospect that I was too early to discount postmodernism. The basic argument of Supermodernism was that after postmodernism and its effete offshoot deconstructivism, another kind of architecture was emerging, which distanced itself from postmodernism. This supermodernism was indifferent to its surroundings, instead of contextual; it prioritized experience, instead of meaning; it promoted a neutral abstraction, instead of symbolic forms and gestures. Simply put, If the museum was the postmodern building type par excellence, the airport was its supermodern equal.

The abstract objects made since 1990 by OMA, Herzog and De Meuron, Toyo Ito, Jean Nouvel, and many others all fitted in the post-postmodern tendency which I called Supermodernism, borrowing from Marc Augé's elegant little book Non-Place: An Introduction to Supermodernity.

My claim that supermodernism had replaced postmodernism was premature, and in hindsight it seems that supermodernism was perhaps just another expression of postmodernism, still driven by postmodern ideas or the negation thereof.

主义的另一种表现形式，仍被后现代思想或其反思想所驱动。

此外，自世纪之交以来，后现代主义经历了一次重大的复兴，形成了新后现代主义建筑风格，包括亚当·纳撒尼尔·福尔曼壮观的建筑作品，以及参加2017年第二届芝加哥建筑双年展"创造新历史"的设计师精心重新审视建筑历史的建筑作品。此次双年展由约翰斯顿·马克李建筑事务所的建筑师莎朗·约翰斯顿和马克·李组织。

但是，正如新哥特式建筑风格既不是对哥特风格的复制，也不是哥特风格的无缝延续，这种新后现代主义，即使在产生时间上更接近它的原身后现代主义，它也并不等同于后现代主义。

不同的是，如今的新后现代主义者对现代主义持有放松平和的态度，而在过去，最早期的后现代主义者经常会感到焦虑和痛苦。后现代主义对现代主义轻蔑和反对的态度已然不见，现在的新后现代主义建筑比以往任何时候都更具包容性，这也体现在地理的包容性上，这最终将促成真正的全球建筑的开端，以及就此形成的全球建筑史。

即使后现代主义仍然存在，它所提出的超现代主义是一种因全球化而产生并侵蚀场所感的建筑形式的反映这一假设，或许比我在1998年写这本书时所理解的更加真实和准确。

不同经济、文化和社会之间的联系越来越密切，因此我们也越来越意识到是全球化将我们带入了如今的世界，在这个世界中，不但我们的场所感被侵蚀了，连这个世界本身、我们的栖息地、我们作为一个物种的存亡以及数以百万计的其他形式的生命都处于危险的境地。

In addition, since the turn of the century postmodernism underwent a serious revival, and right now there is a broad neo-postmodern spectrum, ranging from the spectacular work of Adam Nathaniel Furman to the careful reexamination of architecture's past in the work of many of the designers who were participating in the second Chicago Architecture Biennial in 2017, Make New History, curated by Johnston Marklee's Sharon Johnston and Mark Lee.
But just as neo-gothic was neither a repetition of gothic nor its seamless continuation, this neo-postmodernism, even if it is in time much closer to its original incarnation, is not the same as postmodernism.
What is different is for instance that today's neo-postmodernists are completely relaxed with modernism, which often caused a lot of agitation and distress among the original postmodernists. The disparaging anti-modern attitude of postmodernism is gone, and the current neo-postmodernism architecture is definitely more inclusive than ever, also geographically, finally making it possible to see the beginning of a truly global architecture, and a global architectural history for that matter as well.
Even if postmodernism is still around, the assumption that supermodernism is a reflection in built form of an eroding sense of place, due to globalization, is perhaps more true that I was able to comprehend when I wrote it in 1998.
The increasing interconnectedness of economies, cultures, and societies is coinciding with a growing awareness that this globalization has brought us where we are now: in a world where not only a sense of place is eroded, but place itself as well now that our habitat, our existence as a species and millions of other forms of life is endangered. Instead of the freedom promised by globalization, we are increasingly facing the consequences of the ecological

我们越来越多面临着的，是自工业革命开始以来人类所造成的生态灾难的后果，而不是全球化所承诺的自由。

我在1998年出版的《超现代主义》一书第一版中，表现出了一种漠视一切的乐观主义；在"9·11"事件之后出版的第二版的介绍中，我们无法再漠视全球化的负面影响。但这一切都发生在我们真正进入数字时代之前。自超现代主义早期以来，这一观念发生了巨大变化，现在全球化的概念已经与数字技术完全交融在一起，这些技术不仅模糊了时间和地域上的差异，还常常完全消除了这些差异。

马克·奥格对场所和非场所进行了区分，但这种二元结构正在消失。从定义的场所来看，世界变得越来越难以理解，因为这里和那里、内部和外部、中心和外围，甚至现在和彼时之间的对立，都没有以前那么重要了。

如果说，"万般皆可"是后现代主义的信条，那么现在这一信条是否更接近于因为Wi-Fi的存在，万事在任何时间任何地点皆可？

技术正在消除人类的位置感，而生态灾难可能正在消灭人类的家园。这两种截然不同的体验之间却有着相似之处，让人很难忽略。

如果说在过去，全球化的超现代主义时代是由"世界可以由任何人掌控"这样乐观的想法所主导的，那么现在我们已经意识到，我们都把世界扛在了自己的肩上。

在全球气候变化和环境破坏的背景下，任何关于后现代主义、超现代主义和新后现代主义建筑之间细微差别的讨论，都显得无关紧要。就好像讨论有多少天使可以在一根大头针上同时起舞一样无足轻重。显然我们还有更紧要的事情

disaster humans have caused since the beginning of the Industrial Revolution.
The first edition of Supermodernism which came out in 1998 reflects an unconcerned optimism; in the introduction of the second edition, which was published after 9/11, it was no longer possible not to see the other side of this happy globalization. But this was all still before we truly entered the digital era. What has changed dramatically since the early days of supermodernism as well is that the notion of globalization is now completely entangled with all the digital technologies which have not simply blurred differences in time, and place, but often completely eliminated them.
Marc Augé made a distinction between places and non-places, but this binary is disappearing. The world is less and less understandable in terms of defined places because the opposition between here and there, between inside and outside, between center and periphery, and even between now and then is of less relevance than before.
If one of postmodernism's creeds was "anything goes", it is now perhaps closer to anything goes, all the time, and everywhere, courtesy to Wi-Fi.
It is hard to miss the equivalence between the two contrasting experiences, that technology is eliminating humans' sense of place, and the ecological disaster which is potentially eliminating humans' place.
If the Supermodern period of globalization was driven by an optimistic idea that the world could be anyone's oyster, now the realization has sunken in that we all carry the world on our shoulders.
In the light of the global climate change and the environmental destruction, any discussion about the nuances between postmodern, supermodern and neo-postmodern architecture is as relevant as the question how many angels can dance on the head of a pin. There are obviously more urgent matters at stake. Yet if one wants to make sense of the world, the issue is not only to face what is in front of us, but also to find a position from where to see

要去做。不过,如果我们想要理解这个世界,我们就不仅要面对眼前的问题,同时也要找准看待问题的角度。建筑既可以是一个镜头,也可以是一面镜子,从中我们可以窥探世界,也可以看清自己。

我们可以选择元现代主义视角。元现代主义如今正逐渐崭露头角,简单地说,它旨在揭示所有内部矛盾,并克服后现代主义解构一切的倾向。与后现代主义相反,元现代主义试图将其研究对象在整个领域内进行重建和整合,不否认其存在内部矛盾,但同时又不认为这些矛盾是无法解决的。后现代视角需要有一个激光般锐利的焦点来揭露宏大叙事的矫饰,元现代主义视角则需要一个广角视角以新的方式定位主题,将看似孤立的时刻与所有可能的冲突和矛盾连接起来。元现代主义并不是现代大叙事的回归,而是现代与后现代观念的结合,它不认可只有微观叙事这一种方式,而是认为不只有一种方式来叙述故事。

如果说现代主义建筑是严肃的,后现代主义是讽刺的,超现代主义是冷漠的,那么元现代主义就是严肃与滑稽并存、讽刺与真挚并存、冷漠与热络并存、情感与理智并存、现实与理想并存、无望与希望并存。

在今天这个时代,我们唯一能做的就是,去理解建筑既是一种破坏性的行为,也能够给人们提供舒适的庇护之所;它既能使世界成为一个更美好的地方,同时也不可避免地会危害地球的未来。像我们如今这样继续大兴土木是不可行的,但完全不建造房屋也是不可能的。

因此,元现代主义或许可以为我们提供一个临时的解决方案,能够让我们勉强接受如今的生存状态,而后现代主义的讽刺和超现代主义的冷漠都无法给我们带来任何希望。

it, and architecture can be a lens, and a mirror at the same time to look at the world, and ourselves.
One possible position is to take a metamodern viewpoint. Metamodernism is a slowly emerging set of ideas, which simply put, aims to overcome postmodernism's tendency to deconstruct everything, in order to lay bare all internal contradictions. By contrast, metamodernism attempts to reconstruct and integrate the subject of study in its total field, without denying its inner contradictions, but without accepting these contradictions as unresolvable facts. Whereas a postmodern perspective requires a laser-sharp focus to unmask the pretensions of a grand narrative, the metamodern perspective asks for a wide-angle view, to situate the subject in new ways, connecting the dots between seemingly isolated moments with all their possible conflicts and contradictions. Metamodernism is not a return to the modern grand narrative as such but rather a combination of modern and postmodern notions, refusing to accept there is nothing but micro narratives, but simultaneously accepting that there is not one way of telling a story.
If modern architecture was serious, postmodernism ironic and supermodernism indifferent, a metamodern attitude is to be serious and funny at the same time, ironic and earnest, indifferent and engaged but also simultaneously sentimental and rational, realistic and idealistic, hopeless and hopeful.
It seems that in this day and age this may be the only viable way, to understand that the making of architecture is both a destructive act, and necessary to give people the comfort of being sheltered; both a tool to make the world a better place and concurrently and unavoidably detrimental to the planet's future. Continuing building like we do is not possible, not building at all neither.
So perhaps metamodernism can be today's provisional intellectual refuge, to come to terms with an existential situation for which neither the irony of postmodernism, nor the indifference of supermodernism will be able to offer any hope.

世界前沿的阿拉伯建筑

Arab Archite

波斯湾地区国家的大都市历经30年的发展,到目前为止,已经成功地获得了国际知名度,但其在建筑或者城市规划领域未能像其他国家的大都市那样引领世界潮流或带来深远影响。这些波斯湾地区的城市一直在追求建筑的体量,竞相打破建筑纪录。这比起打造高质量建筑空间更容易实现,因为建筑的体量可以通过测量数字直观地展示出来。

今天,这些城市正在改变其建筑进化发展路径,尝试全新的、与众不同的建筑形式。这些建筑正更多

The metropolises of the countries facing the Persian Gulf have undergone a thirty-year development which has, so far, decreed their success in terms of global visibility, but which has not yet succeeded in defining or strongly influencing the world path of architecture or of urban planning, as has been the case in several cities of other countries.
Gigantism has been pursued in, a race to break building records – a goal much easier to achieve than that of creating a quality architectural space – because the bigness of a building can communicate more easy to the persons, through measurements and numbers.
Today, these cities are searching for new and different architectural forms which could change

珍珠场地，博物馆及珍珠之路入口_Pearling Site, Museum and Entrance to the Pearling Path / Valerio Olgiati
迪拜购物中心苹果店_Apple Dubai Mall / Foster + Partners
卡塔尔国家博物馆_National Museum of Qatar / Ateliers Jean Nouvel
卡塔尔国家图书馆_Qatar National Library / OMA
城市进化：世界前沿的阿拉伯建筑_Evolution on the City: Arab Architecture in front of the World / Diego Terna

地借鉴当地特色，试图将阿拉伯建筑拉回到更加精致的轨道，但仍能满足大众旅游的需求。

下面介绍到的这些项目便符合这条发展路径：在很大程度上，这些项目力图搭建阿拉伯文化与西方文化之间的对话，回顾并借鉴欧洲城市建筑特点。或许通过回归过去，我们能够创造出一种新的建筑类型，并能够以此对其他世界大都市的发展变化产生影响。

their evolutionary path. With an architecture more tied to local references, these new projects try to draw Arab architecture back towards a more refined discourse – but still within a mass tourism mechanism.
The projects presented below follow this path: in large part they seek to bring an Arab culture into dialogue with a Western one, retracing, above all, characteristics of the architecture of European cities. Perhaps a new type of architecture can arise from this return to the past, capable of influencing the evolution of other world metropolises.

城市进化：世界前沿的阿拉伯建筑
Evolution on the City: Arab Architecture in front of the World

Diego Terna

首先，坐标网格是一种概念性的推测。尽管它表面上是中立的，但它暗示着对这座岛屿的一种理智规划：它对地形和现存的东西漠不关心，它声称智慧建设比现实更具优越性。

——雷姆·库哈斯，《癫狂纽约》，1978年

自20世纪90年代以来，波斯湾周边国家在建筑方面便不断进行尝试和挑战。首先从数量上来看他们做到了，这些国家打造了最高的摩天大楼、最倾斜的塔楼、最大的建筑……

这就像一场展示各种可能性的肌肉秀，一种向世界展示自我的方式，一条进入全球话语精英圈的途径——而展示肌肉似乎是获得其他国家关注的最快途径。

这段建筑设计之旅并没有引起人们的极大兴趣，特别是考虑到建造这些巨型建筑所耗费的大量资源。这再次印证了一个道理，那就是应用于建筑本体的工程特性并不一定能在建筑空间质量方面产生积极的结果，相反，建筑设计应该更加重视超越空间质量本身的其他特性。

或许我们必须回到1956年，回到由弗兰克·劳埃德·赖特设计的高1609m的摩天大楼伊利诺伊大厦（高度是哈利法塔的两倍），才能在更大的体量中找到一种质的感觉。然而，这座由美国建筑师设计的建筑是一个"单体"，只是大部分都为水平建筑规划的"广亩城市"规划的一部分。它不像阿联酋的摩天大楼那样可以载入建筑史册，它更像一个感叹号，以其特殊性成为周遭环境中浓墨重彩的一笔。

The Grid is, above all, a conceptual speculation. In spite of its apparent neutrality, it implies an intellectual program for the island: in its indifference to topography, to what exists, it claims the superiority of mental construction over reality. - Rem Koolhaas, *Delirious New York*, 1978.

The countries bordering the Persian Gulf have represented, since the 90s, a sort of continuous challenge to architecture. They did it, above all, in quantitative terms: the tallest skyscraper, the most inclined tower, the largest building …

It has been a muscular exhibition of possibilities, a way to present oneself to the world, to enter the elite of global discourse – and a show of muscles seemed the quickest way to get noticed by other countries.

It has been a journey that did not lead to spatial results of considerable interest, especially when compared with the enormous amount of resources employed in the construction of these gigantic building apparatuses. It is a path that has reaffirmed that the engineering peculiarities applied to the body of architecture do not necessarily lead to positive results with respect to spatial qualities, but instead tend to shift the narration towards other characteristics, which go beyond the quality of the space itself.

Perhaps we must go back to 1956, to *The Illinois* tower, the 1609-meter-high skyscraper by Frank Lloyd Wright (twice the height of the *Burj Khalifa*), to find a qualitative sense in a dimensional quantity. The building of the American architect, however, was a "unicum", which acquired its meaning within a mostly horizontal plan (the *Broadacre City*). It was not an object to be added to a catalog, like the skyscrapers of the Emirates, but rather an exclamation which, in its exceptionality, placed an accent on its surroundings.

Neither is this collection of Arab buildings comparable to organic urban complexes, as could be said of the small Tuscan cities, such as San Gimignano, where the race to build ever taller individual towers was

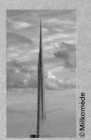
伊利诺伊大厦3D模型，
弗兰克·劳埃德·赖特，1956年
3D model of The Illinois Tower,
Frank Lloyd Wright, 1956

哈利法塔阿联酋
Burj Khalifa, UAE

圣吉米尼亚诺，意大利托斯卡纳
San Gimignano, Tuscany, Italy

　　这些阿拉伯建筑也无法与有机的城市综合体相提并论，即便是在同样开展了高楼竞赛的托斯卡纳小城市圣吉米尼亚诺，设计师们也能够从人体尺度上定义城市空间。

　　纽约从19世纪末便开始进行微不可察的建筑规划（或者说，休·费里斯能在几乎没有任何参考的情况下想象出一座抽象的、不带有人类感情的大都市，这本身就是一种强有力的暗示）。但即便是纽约，其建筑发展路线也与中东地区大都市过去三十年的发展大相径庭。

　　波斯湾地区国家有别于其他地区之处可能就在于建筑沉淀的时间和城市设计方面的不同：在波斯湾，高楼大厦骤然拔地而起，就像一场失控的爆炸。

　　也许更有趣的是，这些波斯湾地区国家将地球上一些最不宜居的地方改造成了城市：他们从无到有，建造了城市以及其发展所需的基础设施系统，填海造田，发展能够自力更生的技术，在恶劣的环境中寻求生存。

　　设计师们打破了沙漠这一单调的现实形态，设想出了另一种地域风貌。简而言之，沙漠已经变成了一座人造绿洲，能够为居民和游客提供比自然条件下的沙漠更为复杂丰富的体验。从这个意义上来说，建筑始终都是实现景观建设和地域多样性的媒介。如果没有建筑，景观建设和地域多样性也就不复存在了。

　　四十多年前，雷姆·库哈斯在纽约的坐标网格中发现了"理智猜测优于物理现实"，我们周围的世界在这一刻解体，取而代之的是一种激进的人为建筑方法：我们周围的土地变成了虚无抽象之地，需要我们用强大的、极度亢奋的姿态来面对。

nevertheless able to define urban space on a human scale.

Not even New York – which from the end of the 1800s was constructed through imperceptible planning (or with so few guidelines that Hugh Ferriss was able to imagine a metropolis of abstract, inhuman, presences – itself a very powerful suggestion) – manages to bring back to a recognizable process, the path made by the Middle Eastern metropolises in the last thirty years.

The fundamental difference is probably the time of sedimentation of architecture and urban design: here it has arrived unexpectedly, like an uncontrolled explosion.

Perhaps more interesting then, are the territorial changes made in some of the least habitable places on the planet: the creation of metropolises from scratch, the infrastructural systems necessary for their development, the construction of new land stolen from the sea, the development of self-reliance techniques to survive in an otherwise hostile environment.

An alternative territory has been imagined, which would allow it to break the monotony of its reality, the desert. In short, it has been transformed into an artificial oasis, which has been able to donate, to its inhabitants and visitors, a much greater complexity than the one it could offer naturally. In this sense, architecture has been the medium for the construction of a landscape, of a territorial variety that otherwise could not have existed.

More than forty years ago, Rem Koolhaas found in the New York grid the realization of a superiority of intellectual speculation over physical reality, a moment of disintegration of the world around us, in favor of a radical anthropic approach: the territory that surrounds us becomes the place of nothingness, of abstraction, to be faced through a strong, delirious gesture.

Here, the desert – the abstract place par excellence, even more so than the homogeneous grid of

毕尔巴鄂的古根海姆博物馆，西班牙
Guggenheim Museum, Bilbao, Spain

  沙漠，作为比曼哈顿单一的网格更为抽象的地方，在波斯湾为新型建筑的发展提供了基础。最重要的是，它还能提供一座新型城市，前提是有理智的设计推测的支持。

  然而，或许我们自己还没有意识到这一点，但我们已经在追求一种未来理念：室内城市，即建筑内部连绵不绝的人文景观。巨大的水族馆、游乐园、滑雪场这些典型的户外元素，都融入了建筑之中，使得室内与室外空间的过渡变得模糊。

  正是建筑的维度使得这一过程得以实现，使得建筑之间的外部空间，即数百年来我们称之为城市的公共空间，变得无用。因此，位于波斯湾地区的国家所构想出的大都市都是标志性建筑的聚集区，而这种建筑的聚集并不等同于一个连贯通透的城市有机体。

  重要的是，人们处于建筑内部的时候，就像在室外一样……而真实的外部世界则变成了一个入口，一条回到一个新的室内空间的快速通道。

  如果我们一直采用这种激进的建筑设计思路，那我们未来的城市很可能与不断扩张的中东大都市如出一辙。

  但事实上，新近设计规划的阿拉伯城市似乎正在回归到更为经典的城市设计轨道，其有以下三个主要原则：

—— 呼吁在建筑实践中更注重建筑的空间品质；

—— 为大都市增添有价值的公共设施；

—— 将一些传统建筑主题融入新建筑。

Manhattan – could offer the basis for a new type of architecture. Above all it could offer a new type of city, provided that it is supported by an intellectual form of design speculation.

However, perhaps without overtly realizing it, an idea of the future has been pursued: the city of interiors, a continuous succession of anthropic landscapes enclosed within buildings. The huge aquariums, the amusement parks, the ski slopes, are all typical outdoor elements, which have been incorporated into architecture, making the transition between what is inside and what is outside indistinguishable.

It is the dimension of architecture that has allowed this process, rendering the external space between the buildings useless – what for hundreds of years we have known as The City, that is, its public space. For this reason the metropolis imagined in the Persian Gulf countries is a collection of iconic architecture, which does not necessarily equate to a coherent, pleasantly permeable organism.

The important thing is to be inside, as if one was outside... while the real outside becomes a threshold, a fast passage, before going back into a new interior.

If this hypothesis were actually pursued with radicality and coherence, the city of the future could well be one of the Middle Eastern metropolises which are constantly expanding.

But in fact, the latest evolution of Arab urban design seems to be trying to return to a more upon classical vision of the city, through three main principles:

- to call upon architectural practices more focused on the spatial qualities of architecture;
- to insert valuable public functions into the metropolis;
- to incorporate some traditional construction themes into new architecture.

They have chosen a more well-known path, the one of a European city, trying to retrace hypotheses already successfully tested (the Guggenheim Museum in Bilbao, for example). They offer visitors the typical places of

卢浮宫与金字塔，巴黎
The Louvre Palace and the pyramid, Paris

他们选择了一条更为人所熟知的道路，即模仿欧洲，学习欧洲城市成功的设计经验（例如，毕尔巴鄂的古根海姆博物馆）。这些建筑为游客提供了典型的全球旅游场所：建筑空间与大众文化密切相关，例如，博物馆或大型单一品牌商店，将建筑设计的兴趣重心从建造庞然大物转向打造更精致的有价值建筑开发项目，通过提升建筑的空间品质来吸引全世界的注意。

这里所展示的项目是阿拉伯大都市新发展的一部分，这些城市希望抛开石油这种黑黄金的支持，能够在国际舞台上占据一席之地；这些城市，就像巴黎、纽约和北京一样，正在成为游客们必去的旅游目的地。从这个意义上来说，如果能在波斯湾地区国家建立起像卢浮宫或古根海姆这样的文化品牌，其将成为大众旅游不可或缺的景点。

因此，要做到这一点，我们必须向游客呈现出代表地域文化的新元素，这里介绍的项目致力于研究当地的建筑类型和技术、摆脱全球特征（至少部分地）、突显当地地理和气候的独特之处。

这里特别要提到瓦莱里奥·奥尔吉亚蒂设计的珍珠场地，博物馆及珍珠之路入口（16页）。该建筑的通道设计力图追求更深的阴影，使游客得以躲避当地炎热的气候。这种阴影设计是通过一座环保的小规模建筑实现的，该建筑打造了一个连接建筑场地各个元素的大型遮蔽物，由一片柱林所支撑。这片柱林是人造景观，但给人一种自然的感觉。其中一些柱子成为风塔，风塔是当地典型的建筑元素，能够在外界极端炎热的情况下产生新鲜的气流。

该建筑的外观使用了古老原始的元素，在这位瑞士建筑师的项目中尤为突出：和波斯湾地区其他建筑不同，建筑师并没有将建筑内部设计得如同外部一样，相反，随着建筑所在地城市化规模的扩大，该建筑试图寻求与周围环境建立联系。

global tourism: spaces linked to mass culture, such as museums, or large single-brand stores, shifting the axis of interest from engineering gigantism to a more refined development of valuable architecture, capable of capturing the attention of the world by their spatial qualities.

The projects presented here are part of the new evolution of Arab metropolises, cities that want to impose themselves on the international scene beyond the support of *black gold*; they are becoming places of necessary visit, just as Paris, New York, Beijing. In this sense, the possibility of bringing cultural brands like the Louvre or the Guggenheim here, create indispensable points of visit in mass tourism journeys.

To do this, therefore, it is necessary to present to visitors new elements that represent the culture of the place: the projects presented here work on discovering local typologies and technologies, freeing themselves from global characteristics (at least in part) and highlighting the unique idiosyncrasies of the place and climate.

The Pearling Site, museum and entrance to the Pearling Path (p.16) by Valerio Olgiati, in particular, defines a distribution path that is based on an exasperated search for deep shadows, allowing visitors to be sheltered from the torrid local climate. It does so through a low-impact, small-scale building that builds a large shelter connecting the various elements of the site, supported by a forest of pillars, an artificial landscape that crystallizes a natural sensation. Some of these pillars then become wind towers, typical elements of local architecture, able to create currents of fresh air despite the external heat.

The exterior is the original, yet ancient, element that is highlighted in the Swiss architect's project: an interior in the form of an exterior is not created – as in many buildings of the area – but rather a relationship with the surrounding territory is sought, with the urban scale enclosing the new intervention.

The only concession to a more international approach is the form, the continuous exhibition of a geometric

卡塔尔国家博物馆
National Museum of Qatar

该建筑在设计上唯一称得上国际化的方面就是它的形式,整座建筑中都充斥着连续不断的几何图形。这些几何图形是这座建筑的标志,矩形结构上覆盖着坡屋顶,这样的屋顶设计更进一步缩小了建筑的规模,使建筑更加迎合游客的身体和视觉需求。

福斯特建筑事务所设计的迪拜购物中心苹果店(30页)是全球旅游购物圣地之一,该建筑前所未有地将外部空间囊入了大型购物中心之内。其大型洞口设计既避免了建筑的完全闭合,又支持了室内外空间之间的交互。这些洞口能够旋转,会在晚上将朝向立面的空间全部打开;而在白天,这些洞口可以调节阳光,将商店内部照亮。

这是对传统阿拉伯窗花的一种技术应用,从这个意义上说,该建筑工程解决方案的目的并不是追求建筑的体量,而是重拾阿拉伯地方特色。苹果公司这个通过其产品改变世界文化的国际品牌,在这里发现了另一种发展进化模式,也因此吸引了大批游客。

由让·努维尔工作室设计的卡塔尔国家博物馆(40页)将象征的手法融入建筑设计之中:该博物馆如同一朵巨大的沙漠玫瑰(一种沙漠中的沉积结构)。用这位法国建筑师的话说:"沙漠玫瑰……是千百年来,大自然在风、海浪和沙粒的共同作用下,创造出的第一个建筑结构。它出人意料地复杂并富有诗意。"

从小小的矿物结构沙漠玫瑰,到大型的博物馆建筑,随着建筑规模的扩大,建筑师也面临着技术挑战,需要很多资源来精确构造出建筑的初始图像,甚至在这个建筑实践中,建筑师还使用了大块金属板来控制太阳辐射,在建筑立面上创造了一系列较深的阴影。建筑师还利用这些金属板打造出完全遮阳的空间,创造出了无数的凉爽微气候,改善了整座建筑的环境。

figure that instructs the entire project. It is the icon of the house, the rectangle surmounted by a pitched roof, which reduces the scale of the project even more, approaching the physical and visual sensations of visitors.

The Apple Dubai Mall (p.30) by Foster + Partners (one of the secular cathedrals of global tourism) tries, in an unprecedented manner, to let the outside penetrate inside a large shopping center. It avoids the complete closure of the architecture, but it favors exchange between the two worlds. It does this through large openings that, by rotating, completely open the spaces towards the facade, in the evening. During the day, these allow a modulation of the sunlight to enter the store.

This is a technological application of the traditional *Arabic Mashrabiya*: in this sense the engineering solution is not developed to give shape to a quantity, a dimension, but to the recovery of a local element. An international brand, capable of changing world culture through its devices, finds here a different evolution and for this reason it becomes attractive for mass tourism.

The National Museum of Qatar (p.40) by Ateliers Jean Nouvel applies a rhetorical form to the construction: it builds the museum like a giant desert rose, a sedimentary formation found in the desert. In the words of the French architect: "The desert rose [...] is the first architectural structure that nature itself creates, through wind, sea spray and sand acting together over millennia. It's surprisingly complex and poetic."

The scale enlargement, from a small mineral to a large building, thus becomes a technological challenge in which resources are used to precisely construct the initial image: even in this case, the large plates become useful for controlling the solar radiation, to create a series of deep shadows on the facade of the building. They also define spaces that are completely sheltered from the sun, giving rise to numerous cooler

卡塔尔国家图书馆
Qatar National Library

    这里要介绍的最后一个项目是由OMA设计的卡塔尔国家图书馆(62页),该建筑的设计思路与前文提到的项目略有不同:或许是基于对纽约项目的分析,库哈斯这次没有直接以当地建筑作为参考。当然,他在建筑设计中运用到了文化元素,他所设计的图书馆既是一个新的旅游目的地,又是一个大型的城市服务机构,但他这种建筑设计依托于阿拉伯大都市在过去30年的发展。现在我们拥有了强大的建筑机器,能够建造极其复杂的室内宏观世界。但是,即使在这个项目中,我们也仍然能够在建筑中发现明显的文化元素,而这也确实成为该建筑的核心,尽管这种对当地文化特色的参考是以一种亲密、雅致的形式呈现的:图书馆中那些只有专家才能借阅的最古老的书,被埋在地面层之下6m深的地方,但这些书从上方依然可见,就像考古发掘一样,而读者的借阅活动也与考古发掘如出一辙。

    环顾四周,大量的藏书构成了一个新的景观,给了书籍这一新的群体以生命。这些书籍是当地居民的伙伴,图书馆室内变成了一个神圣的地方,一个通过文化媒介朝圣的地方。

    OMA设计的图书馆也许表明了阿拉伯城市的转型之路,与奥尔吉亚蒂的珍珠之路项目相对却又互补:前者象征着超级室内、巨型建筑内部人文景观的演变;而后者则是一个更加具有人体尺度的城市,在室内和室外、传统和现代之间建立了更紧密的对话。

    从两个建筑案例中可以看出,这些阿拉伯国家的建筑和城市规划似乎都处于转折点,也许能够影响其他世界大都市的发展和演变。

microclimates, which improve the climate for the entire structure.
The last project presented, the Qatar National Library (p.62) designed by OMA, differs slightly from this path: perhaps mindful of the work done on the analysis of New York, Koolhaas avoids a direct reliance on local architecture. He certainly designs a cultural element, a new tourist destination as well as a large-scale service for the city, but he does so in the wake of the Arab metropolis as it has been evolving over the last thirty years: with powerful architectural machines, capable of creating internal macrocosms of great complexity. And yet, even in this case, the reference to the local is clear, indeed it becomes the core of the project, albeit in an intimate, refined, form: the oldest books in the library, which can only be consulted by experts, are buried six meters under the general walking surface, but they remain visible from above, like archaeological excavations, as do the activities of their readers.
All around, the huge collection of books creates a new landscape, giving life to a new population, the one of the books. These are companions to the local inhabitants: the interior is transformed into a sacred place, a place of pilgrimage via a cultural medium.
The library by OMA perhaps indicates a path of transformation of the Arab city, opposite yet complementary to the project by Olgiati: one is an evolution of the Super Interiors, of the anthropic landscapes enclosed within gigantic building bodies; the other is a more human-sized city, with a closer dialogue between inside and outside, between tradition and modernity.
In both cases it seems that the architecture and urban planning of these Arab countries could be at a turning point, perhaps capable of influencing the growth and evolution of other world metropolises.

珍珠场地，博物馆及珍珠之路入口
Pearling Site, Museum and Entrance to the Pearling Path
Valerio Olgiati

瓦莱里奥·奥尔吉亚蒂在巴林联合国教科文组织文化遗产珍珠之路的入口，
为穆哈拉克设计了一个城市空间
Valerio Olgiati designs an urban room for Muharraq at the Entrance to the Pearling Path,
Bahrain's UNESCO heritage trail

该项目的场地位于巴林的穆哈拉克，包含了一片建筑遗迹，是联合国教科文组织文化遗产珍珠之路的一部分。珍珠之路全长3.5km，连接着三个近海牡蛎养殖场和17座与珍珠产业相关的重要历史建筑，这其中包括布马希尔堡，以及富商住宅、商店、仓库和一座清真寺。新建筑类似于一个巨大的红色混凝土雨篷，整座建筑被作为文化遗址的入口，同时也是进入麦地那市的门厅，与一座博物馆和一个公共空间结合在一起。

该建筑旨在为穆哈拉克人民提供一个与城市公园规模相当的城市空间，围绕在场地边缘的混凝土体块在拥挤的城市中构建了一个新的中央场所。柱林和风塔支撑着距地面10m的水平屋顶，形成了一个巨大的空间。用瑞士建筑师瓦莱里奥·奥尔吉亚蒂的话来说："屋顶呈古典样式，在炎热的气候下为市民们带来难得的阴凉，并通过其超大的规模营造出一种独特的氛围。"

建筑顶部粗壮的柱子构成了捕风器，这些柱子贯穿了雨篷的屋顶。这些柱子能为建筑降温，同时柱子底部的壁龛能够遮挡建筑座位区，为其提供阴凉。雨篷遮挡着"阿玛拉"废墟，这片废墟最初是仓库、工厂和市场。稍往后一点，一座带有神秘色彩的房子藏于阴影之中，联合国教科文组织文化遗产博物馆便坐落于此。整座建筑本身便是一个宇宙，是珍珠之路乃至更广阔的城市的入口。

The site in Muharraq, Bahrain, contains ruins that form part of the UNESCO Pearling Path – a 3.5km trail which connects three offshore oyster beds with seventeen Muharraq's key historic buildings associated with the pearling industry.

These include the Qal'at Bu Mahir fortress, and the residences of wealthy merchants, shops, storehouses and a mosque. The new building – akin to a huge, red, concrete canopy – functions in its entirety as the entrance to one such cultural heritage site, and the foyer for the Medina, combining a museum with a public space.

It is intended as an urban room for the people of Muharraq, with the scale of a public park; concrete elements are placed along the property boundary to form a new locus in a dense city. In the large space, a forest of columns and wind towers hold a horizontal plate 10 meters above the ground. In the words of Swiss architect Valerio Olgiati: "A roof, understood as an archaic gesture, donates vital shadows for the people of Muharraq in this very hot climate, and produces a new and unique situation through its different scale."

On the top, wind catchers are formed by thick columns which perforate through the roof of the canopy. These help cool the structure, whilst niches in the base of the columns create sheltered seating areas. The canopy covers the ruins of an "amara" – the entity which originally functioned as warehouse, factory and marketplace. Slightly set back, in the shadows, is the enigmatic house in which the museum of the UNESCO World Cultural Heritage is located. As a totality the building creates a universe in itself that is the entrance for the Pearling Path and the city beyond.

屋顶 roof

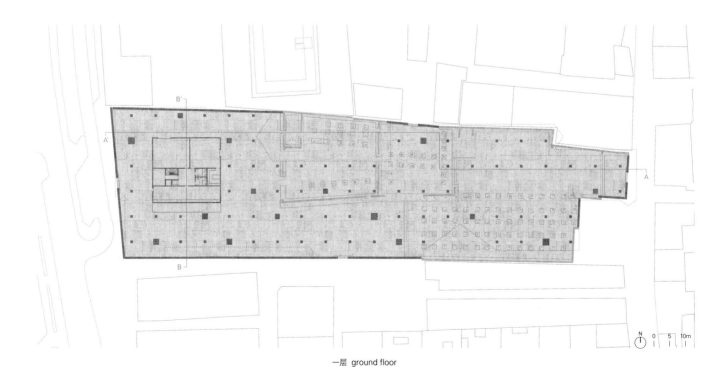

一层 ground floor

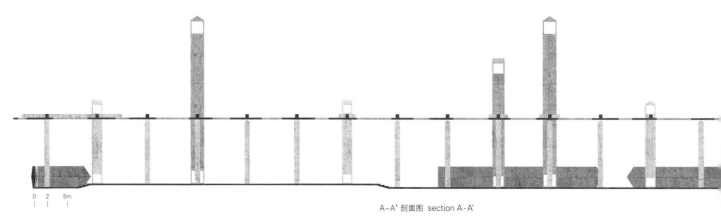

A-A' 剖面图 section A-A'

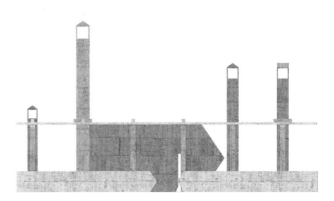

西立面 west elevation

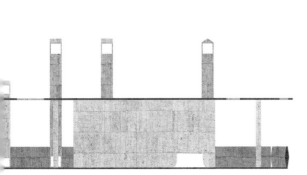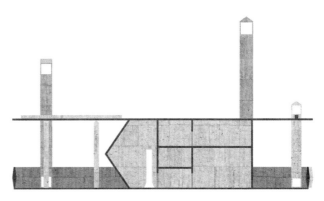

B-B' 剖面图 section B-B'

项目名称：Pearling Site, Museum and Entrance to the Pearling Path
地点：Building 999, Road 10, Block 215, Muharraq, Bahrain
建筑师：Valerio Olgiati
当地事务所：Emaar Engineering
合作者：Sofia Albrigo – project manager; Anthony Bonnici / 总承包商：Almoayyed Contracting Group
客户：Bahrain Authority of Culture & Antiquities / 用途：UNESCO Site museum and entrance / 体积：49,855m³
面积：6,726m² / 材料：In-situ concrete, steel / 规划开始时间：2016.8
施工时间：2018.5—2019.2 / 竣工时间：2019
摄影师：©Archive Olgiati

迪拜购物中心苹果店
Apple Dubai Mall
Foster + Partners

福斯特合伙人事务所在迪拜购物中心苹果店的设计中对阿拉伯窗花进行了重新诠释
Foster + Partners reinterpret the Mashrabiya at the Apple Dubai Mall

迪拜购物中心苹果店改变了传统内向型的商场零售店设计理念，将建筑与城市生活景象结合起来，赋予建筑更具外向感的体验。

建筑师根据阿联酋的文化与气候特征对建筑进行了极具创新性的设计，同时也展示了苹果公司的开拓雄心——打造出一个服务于所有人的、振奋人心的公共空间。

该苹果店位于迪拜购物中心，这里是全世界游客最多的城市中心之一，自2014年以来每年吸引超过8000万名游客。新迪拜购物中心苹果店位于城市最为中心的地段，毗邻地标性建筑哈利法塔，俯瞰著名的迪拜喷泉。店铺共两层，拥有一座喷泉剧院和一个56.6m宽、5.5m深的宽敞露台，这在所有的苹果店中尚属首例。同时该建筑拥有无与伦比的视野，人们置身其中可以看到壮观的景色和令人难以置信的舞台布景。

迪拜购物中心苹果店的设计是一场太阳的庆典，建筑利用丰富的日光营造出别具一格的室内氛围。"太阳之翼"的设计极具创新性，是对传统阿拉伯窗花（一种镶有雕花木制品的凸窗）的重新诠释，它在白天轻柔地为外面的露台遮阳，而在晚上隆重地打开，露出"室内最好的座位"，展现出美到令人窒息的长廊水榭及喷泉景观。

建筑师受到了猎鹰展翅的启发，继而设计出了"太阳之翼"的运动路径。这本身就是一种戏剧性的体验，是一个将运动艺术与工程学结

合起来的构想。翅膀经过精心制作，其外形与功能紧密结合，十分赏心悦目。

每个翅膀都由轻质的碳纤维制成，多层管在翅膀上形成了密集的网。在深入研究了日照角度之后，建筑师在一年中阳光辐射最集中的位置布置了更多的管网。翅膀的花纹设计也很独特，人们站在商店的任何一层，都可以清晰地看到户外的景色；阳光透过翅膀投下斑驳的阴影，一直洒向商店深处。新建的遮阳露台摆放有九盆高树大盆栽，大型花盆结合了供游客栖坐、放松和欣赏风景的座椅。这些花盆会在机械的作用下自动旋转，以保证树木能够均匀地照射到阳光。

Apple Dubai Mall reinvents the traditional, introverted, idea of mall-based retail into a more outward-looking experience that engages with the spectacle of urban life.

Foster + Partner's design is a highly innovative response to the culture and climate of the Emirates, while also demonstrating Apple's pioneering ambition to create inspirational civic spaces for all.

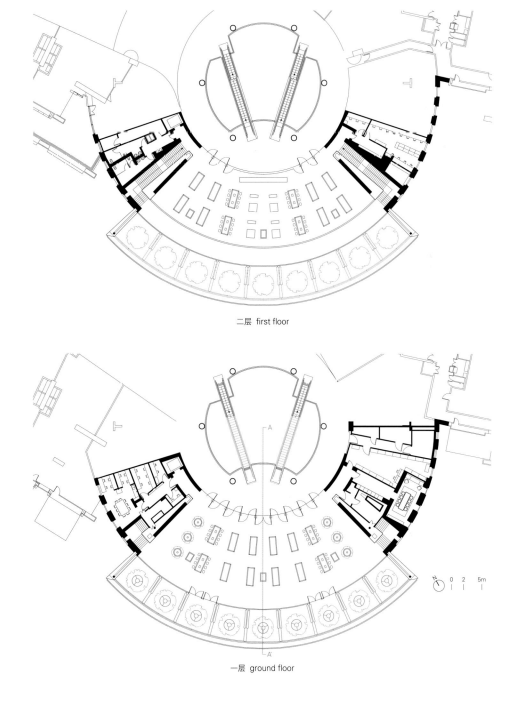

二层 first floor

一层 ground floor

Located in Dubai Mall – one of the most visited urban centers in the world, attracting over 80 million visitors every year since 2014 – the new Apple Dubai Mall occupies the most pivotal position in the city, alongside the iconic Burj Khalifa and overlooking the famous Dubai Fountains. Spanning over two floors, it embraces the theater of the fountains with a sweeping 56.6-meter-wide and 5.5-meter-deep terrace – a first for any Apple Store – with unparalleled views of the spectacular setting and the incredible choreographed display.

The design of Apple Dubai Mall is a celebration of the sun, using the abundant daylight to create a special ambience within. Reinterpreting the traditional Arabic Mashrabiya (a projecting oriel window enclosed with carved wood latticework), innovative "Solar Wings" gently shade the outside terrace during the day and open majestically during the evening to reveal the "best seat in the house" with a breathtaking view of the waterside promenade and fountains.

With their movement path inspired by a falcon spreading its wings, the "Solar Wings" form a theatrical experience – an integrated vision of kinetic art and engineering. The wings have been carefully crafted to delight, a delicate combination of form and function.

Made entirely of lightweight carbon fiber, each wing has multiple layers of tubes forming a dense net. Following an in-depth study of the angles of the sun, the rods have been distributed in higher concentration where the solar radiation is most intense over the year. The unique pattern allows clear views out for people standing on both levels of the store, and the sunlight streaming through the wings casts dappled shadows deep into the interior. The new shaded terrace features nine substantial trees within large planters incorporating seating for visitors to sit, relax and enjoy the view. The planters will rotate mechanically to ensure that the trees receive even sunlight.

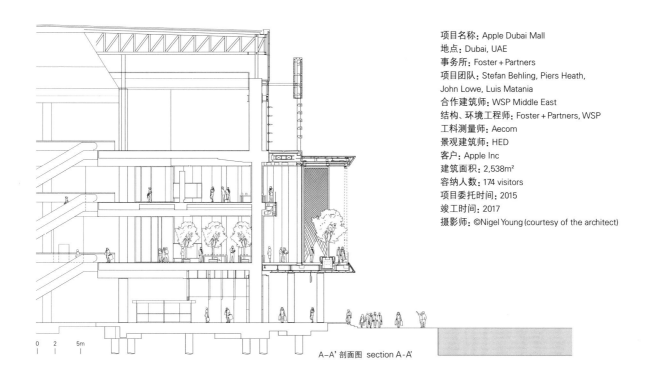

A-A' 剖面图 section A-A'

项目名称：Apple Dubai Mall
地点：Dubai, UAE
事务所：Foster + Partners
项目团队：Stefan Behling, Piers Heath, John Lowe, Luis Matania
合作建筑师：WSP Middle East
结构、环境工程师：Foster + Partners, WSP
工料测量师：Aecom
景观建筑师：HED
客户：Apple Inc
建筑面积：2,538m²
容纳人数：174 visitors
项目委托时间：2015
竣工时间：2017
摄影师：©Nigel Young (courtesy of the architect)

太阳之翼
Solar Wings

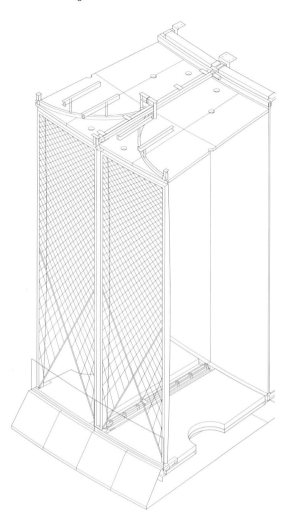

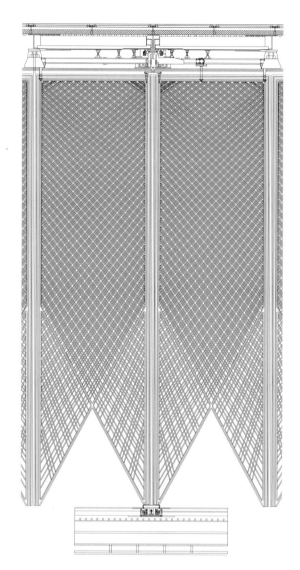

部分立面图 sectional elevation

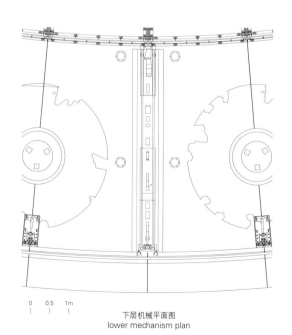

0 0.5 1m

下层机械平面图
lower mechanism plan

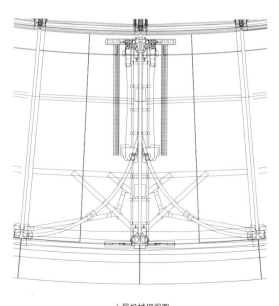

上层机械仰视图
upper mechanism reflected plan

卡塔尔国家博物馆
National Museum of Qatar

Ateliers Jean Nouvel

让·努维尔设计的卡塔尔国家博物馆像沙漠玫瑰般绽放
Jean Nouvel's National Museum of Qatar blossoms like rose in the desert

让·努维尔工作室以沙漠玫瑰矿物结构为灵感来源,设计了新卡塔尔国家博物馆,创造了一座独一无二的、几何形状复杂的建筑。

卡塔尔国家博物馆于2019年3月正式向公众亮相。这座由让·努维尔设计的建筑杰作,为全球访客带来了无与伦比的沉浸式体验。

卡塔尔国家博物馆内的展览长廊蜿蜒曲折,长达1.5km,从建筑空间、音乐、诗歌、口述历史、考古文物、纪念画作、艺术电影等方面,生动演绎着属于卡塔尔的故事。访客穿梭于包罗万象的展馆,仿佛经历了一场旅行。

馆内有11个固定陈列展馆,穿越这些展馆,访客可以了解到从数百万年前卡塔尔半岛的形成,到如今现代多元的国际城市的漫长历史演变。卡塔尔国家博物馆不但展现了这个国家丰富的文化历史和卡塔尔人的美好理想,同时也是一个枢纽,促进了探索、创意、社区活动和教育发展。

正如建筑师所解释的那样,卡塔尔作为"一个吸引着来自四面八方游客的十字路口",它的特色体现在三段不同的历史故事中:半岛游牧民族的宁静生活,以及本土动植物的故事;罗马时期珍珠贸易带来经济发展的故事;20世纪石油和天然气的重大发现,使卡塔尔成为一个繁荣强盛国家的故事。

卡塔尔国家博物馆的核心组成部分是一座经过修复的历史悠久的宫殿,这座宫殿曾属于现代卡塔尔创始人之子Sheikh Abdullah bin Jassim Al Thani殿下(1880—1957年)。这座历史建筑曾是皇家住所、政府所在地,后来被改建成最初的国家博物馆,而现在则成为全新的卡塔尔国家博物馆中最耀目的亮点。

在设计过程中,让·努维尔从海湾地区"沙漠玫瑰"这一神奇的自然现象汲取了灵感,"沙漠玫瑰"是海湾地区一种自然形成的矿物晶体结构,因形似玫瑰而得名。让·努维尔将沙漠玫瑰形容为"第一个由大自然创造的建筑结构"。建筑师以"沙漠玫瑰"为建筑结构模型,创造了复杂的结构,建筑中尺寸和弯曲度不同的圆盘互锁连接,有些是垂直的,起到支撑作用,有些是水平的,错落叠加,这些圆盘如项链一般环绕着历史宫殿。让·努维尔写道:"博物馆占地面积很大。从你走进博物馆的那一刻起,你就会被它形式和规模之间的反差所震撼……被那从时间的迷雾中飘向我们的小小沙漠玫瑰和这个巨大的建筑之间的反差所震撼。至于沙漠,它一直在那里,即使它已经完全变成了另一种形态。"

Baraha中央庭院坐落在展馆中心,是户外文化活动的聚集点;在海湾地区,"豪斯"是一种传统庭院,四周环绕着供游客前来卸货的建筑。外观上,博物馆沙漠色的混凝土结构与沙漠环境相得益彰,建筑仿佛破土而出。在建筑内部,互锁连接的圆盘结构连续设置,形成形式多样的不规则体量。建筑师说:"我们的构想是要创造出一种对比的反差,给人们带来惊喜……这样就创造出了具有动感和张力的建筑形态。"

建筑采用了大量具有可持续性的设计元素,自然遮阳的悬臂式圆盘便是其中之一,这令卡塔尔国家博物馆成为首个获得LEED金奖认证以及全球可持续性评估体系四星级评级的博物馆。

让·努维尔说道:"为了建造这个外形如沙漠玫瑰般充满巨型曲面圆盘、交错设置和悬臂结构的建筑,我们面对了巨大的技术挑战。和卡塔尔这个国家一样,这座建筑采用了大量的前沿技术。因此,它是一个完整的作品,为人们提供了一种融合建筑、空间和感官的独一无二的体验。"

Atelier Jean Nouvel designs new National Museum of Qatar using the desert rose mineral formation as inspiration for a unique and complex geometric architecture.

The National Museum of Qatar (NMoQ) opened to the public in March 2019, welcoming the world to a unique immersive experience designed by Jean Nouvel.

The museum's winding 1.5 km gallery path journeys through a series of encompassing environments, which tell the story of Qatar through architectural spaces, music, poetry, oral histories, archaeological and heritage objects, commissioned artworks, art films, and more.

Eleven permanent galleries take visitors from the formation of the Qatar peninsula millions of years ago to the nation's exciting and diverse present. Expressing Qatar's rich heritage and culture, and the aspirations of its people, NMoQ is a hub for discovery, creativity, community engagement and education. As the architect explains, Qatar's character as "a real cross-

road, alluring and open, attracting visitors from far and wide" is covered in three different stories: that of the tranquil lives of the nomadic people of the peninsula, and its native flora and fauna; the economic development occasioned by the pearl trade in Roman times; and the enormity of the discovery of oil and gas in the 20th century, making Qatar a powerful and prosperous nation.

The museum's centerpiece is the restored historic Palace of Sheikh Abdullah bin Jassim Al Thani (1880-1957), son of the founder of modern Qatar. Formerly the home of the Royal Family, the seat of government, and subsequently the original National Museum, the historic palace is now the culminating exhibit of the gallery experiences.

In designing the building, Jean Nouvel drew inspiration from the desert rose, a flower-like formation of mineral crystals that occurs naturally in the Gulf. Described by Nouvel as "the first architectural structure that nature itself creates," it is the model for a complex structure of large interlocking disks of different diameters and curvatures – some vertical and constituting supports, others horizontal and resting on others – which surround the historic Palace like a necklace. Jean Nouvel writes: "The museum occupies a vast area. From the moment you step inside you're struck by the relationship between the form and the scale … between the small desert rose that comes down to us from out of the mists of time and this outsize creation. As for the desert, it's

always there, even if it has morphed into something else completely."

A central court, the Baraha, serves as a gathering space for outdoor cultural events; in the Gulf, the Housh is a traditional courtyard surrounded by buildings where travelers would come and unload their merchandise. On the outside, the museum's sand-colored concrete harmonizes with the desert environment – the building appears to grow out of the ground. Inside, the structure of interlocking disks continues, creating an extraordinary variety of irregularly shaped volumes. According to the architect: "the idea was to create contrasts, spring surprises … This produces something dynamic, tension."

The building's cantilevered disks, which provide natural shade, are among the elements that have enabled NMoQ to become the first museum to receive both LEED Gold certification and a four-star sustainability rating from the Global Sustainability Assessment System.

Jean Nouvel said, "To construct a building with great curved disks, intersections, and cantilevered angles – the kind of shapes made by a desert rose – we had to meet enormous technical challenges. This building is at the cutting edge of technology, like Qatar itself. As a result, it is a total object: an experience that is at once architectural, spatial, and sensory, with spaces inside that exist nowhere else."

东立面 east elevation

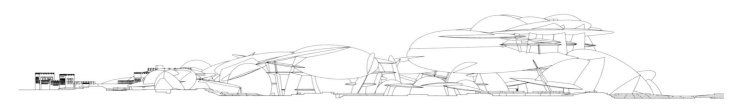

南立面 south elevation

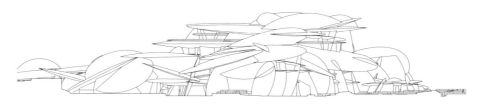

西立面 west elevation

北立面 north elevation

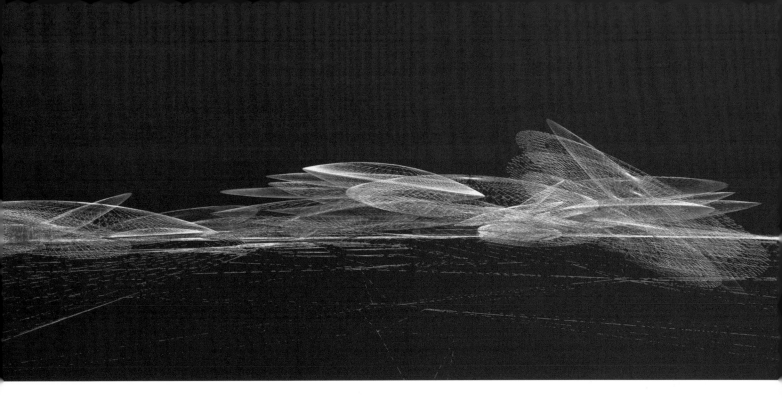

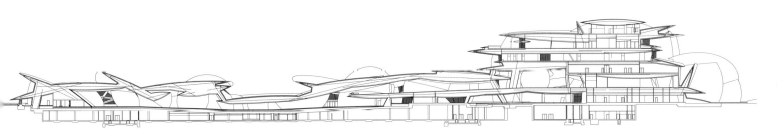

A-A' 剖面图 section A-A'

B-B' 剖面图 section B-B'

C-C' 剖面图 section C-C'

D-D' 剖面图 section D-D'

E-E' 剖面图 section E-E'

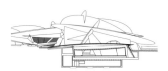

F-F' 剖面图 section F-F'

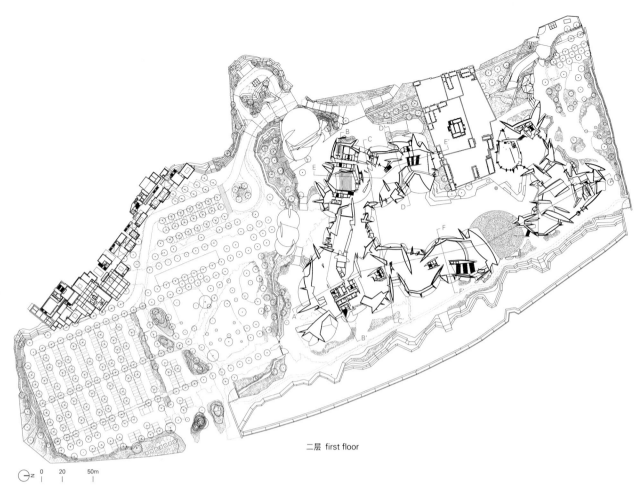
二层 first floor

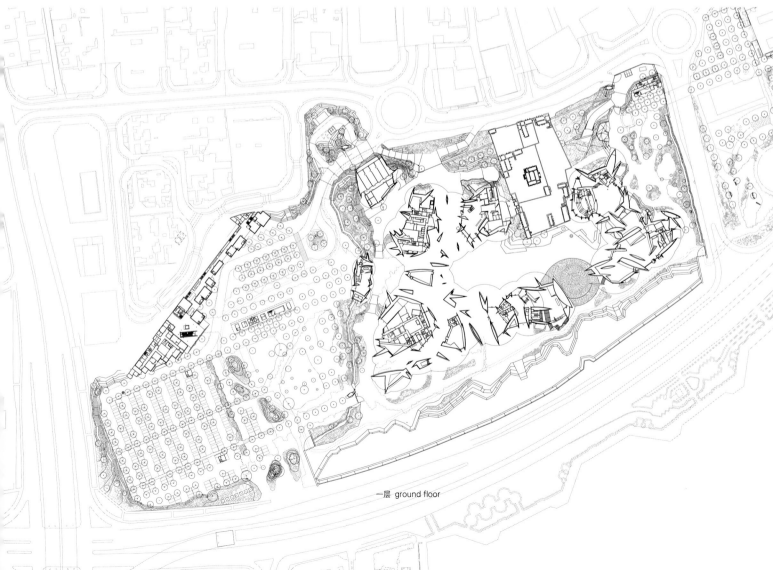
一层 ground floor

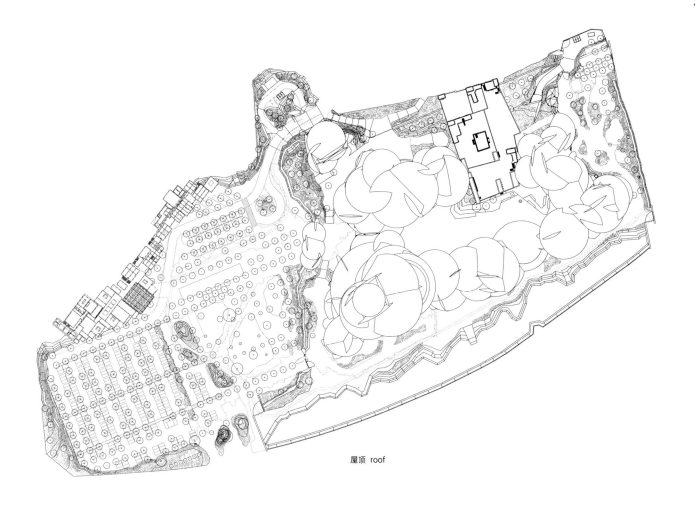

屋顶 roof

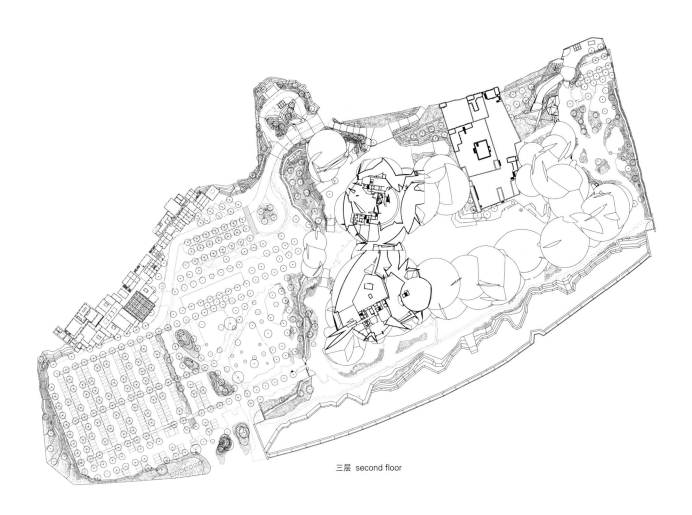

三层 second floor

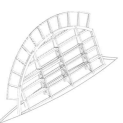

垂直圆盘组成
vertical disk built up

主要结构
primary structure

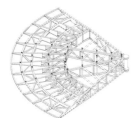

水平圆盘组成
horizontal disk built up

主要结构+末节区域
primary structure + stubs

次级覆层结构
secondary cladding structure

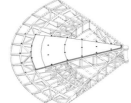

保温与防水复合结构
insulation and waterproofing complex

带状覆板
band of cladding panels

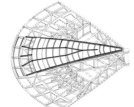

次级覆层结构
secondary cladding structure

最终覆层布置
final cladding arrangement

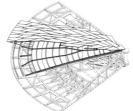

最终覆层布置
final cladding arrangement

互锁圆盘组成
intersection disks built up

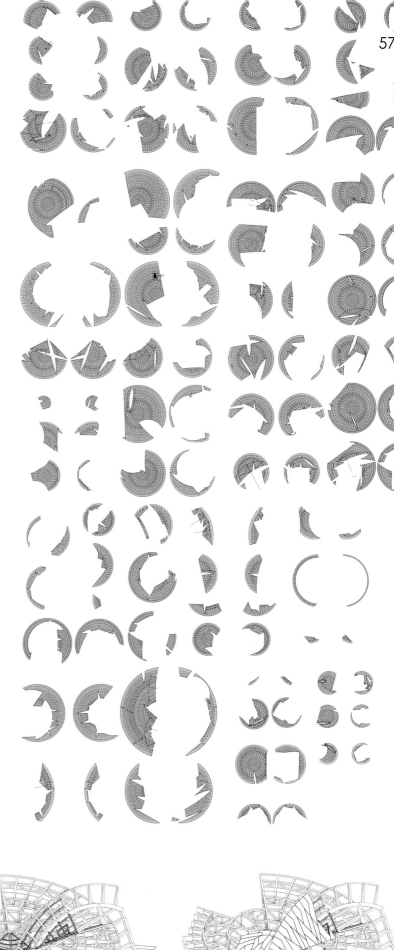

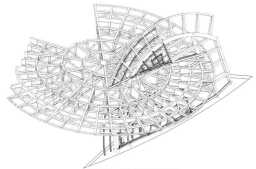

主要结构与垂直圆盘覆层结构
primary structure and vertical disk cladding structure

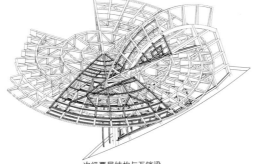

次级覆层结构与互锁梁
secondary cladding structure with intersection beams

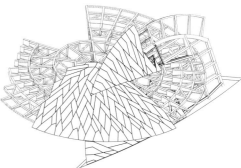

最终覆层布置
final cladding arrangement

The Formation of Qatar	The Archaeology of Qatar	The People of Qatar	Life in Al Barr (Desert)	Pearls and Celebrations	Qatar Today
	Qatar's Natural Environments		Life on the Coast		The Modern History of Qatar

Jila'at Sheikh Abdullah bin Jassim

Museum Baraha

main entrance

项目名称：National Museum of Qatar / 地点：Al Corniche Street, Doha, Qatar / 建筑师：Jean Nouvel - Ateliers Jean Nouvel
项目经理：Hafid Rakem / 总经理：Eric Maria (EMA) / 项目交付经理：Brian WAIT / 建筑师负责人：Philippe Charpiot, Nikola Radovanovic / 质量保障经理：Daniela Fortuna / 项目负责人：Toshihiro Kubota, Eric Stephany / 室内设计负责人：Sabrina Letourneur / 设计顾问：Renaud Pierard / 景观建筑师：Michel Desvigne Paysagiste / 照明设计师：Scherler - interior lighting; LKL - museographic lighting / 客户：Qatar Museums (QM) / 承包商：Mission complète - full design services / 景观面积：112,000m² / 用地面积：143,145m²
总楼面面积：52,167m² / 建筑面积：33,618m² / 可用楼面面积：30,064m² / 研究时间：2003—2011 / 概念设计时间：2008.3
施工时间：2011.9—2018 / 开放时间：2019.3 / 摄影师：©Iwan Baan (courtesy of the architect) - p.40~41, p.44~45, p.46~47, p.50~51, p.52~53, p.56; ©Danica Kus (courtesy of the architect) - p.59, p.60, p.61

卡塔尔国家图书馆
Qatar National Library
OMA

卡塔尔国家图书馆歌颂"书本的活力"
Qatar National Library celebrates the "vitality of the book"

由OMA设计的卡塔尔国家图书馆对于多哈新学术中心的形成起到至关重要的作用,中东地区一些最为罕见的书籍都存于该图书馆建筑的中心。

卡塔尔国家图书馆包括多哈国家图书馆、公共图书馆和大学图书馆,并藏有关于阿拉伯-伊斯兰文化的珍贵文献和手稿遗产。图书馆将收录超过一百万册书籍,并为数千名读者提供超过42 000m²的阅读区域。图书馆周围分布着全世界众多顶尖院校的分校区,在教育城市项目中扮演着重要的角色。该项目由矶崎新于1995年进行总体规划和设计,并于2003年开始施工,建设了一系列教育和研究机构,这其中包括国际知名大学的分校区,以及OMA设计的卡塔尔基金会的总部。

自从赢得1989年法国国家图书馆设计方案竞赛以来,OMA便长期保持着对图书馆设计的兴趣和关注,卡塔尔国家图书馆便是其最新的一项成果。早在那个年代,"电子革命"的兴起便已使实体知识媒介显得不那么重要了,同时也使人们不禁产生了"我们还需要图书馆吗?""数字文化的时代里,图书馆还能够生存吗?"诸如此类的质疑。用该项目建筑师的话来说:"建筑和著书有一个相似之处,那就是二者都有着令人难以置信的悠久传统,但同时又不得不顺应当下,为了生存而不断更新。"

建筑师希望通过卡塔尔国家图书馆这一项目使书籍的活力得以展现。他们希望借助该建筑设计来激发人们学习、研究、协作乃至人与书籍的互动。这一百多万册的藏书中包含着中东地区最为重要且极其罕见的手稿。

图书馆被设计为一个单一的空间,人与书籍同时被容纳其中。建筑的边缘从地面上抬升,形成三条能够存放书籍的走廊,同时围合出一个三角形的中央空间。雷姆·库哈斯说:"这就好像我们拿来一个盘子,把它的边折起来,既创造出了容纳书籍的阶梯平台,同时又留出了进入房间中心的通道。进入图书馆后,你就会立刻被整座图书馆的图

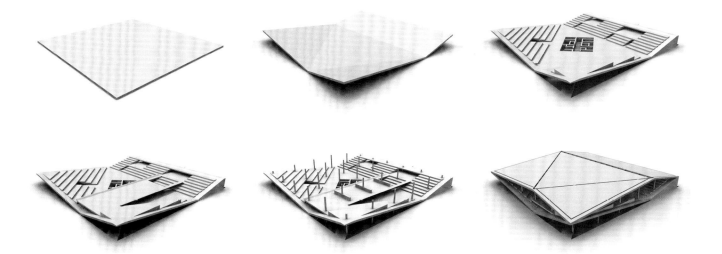

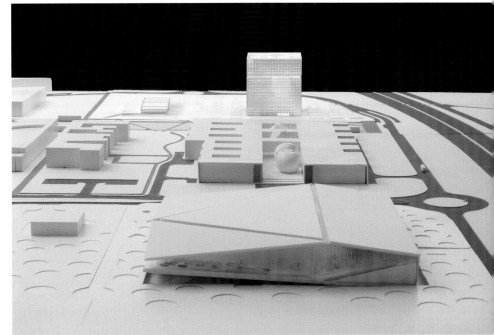

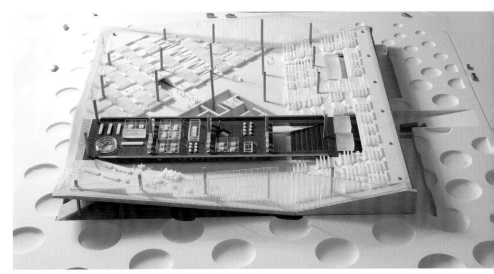

书所包围——所有的书籍都实实在在地存在着，看得见摸得着，读者无须费力便可随手获取。"

在这样的设计中，三条通道延伸于书架之间，沿路分布着阅读、社交和浏览区域。这些书架不论从材质还是从功能方面都与建筑融为一体：其使用了与地面相同的大理石，并结合了人工照明、通风以及还书系统等基础设施功能。

无立柱的桥梁连接了图书馆的主干走廊，开辟出了通向各个区域的路径。这座桥梁上容纳了媒体室和学习室、阅读桌、展览空间以及大型多功能礼堂，由来自阿姆斯特丹的InsideOutside工作室设计的可伸缩的幕帘将其围合其中。

遗产藏书馆位于图书馆中央的一个6m深的空间，这个空间好像是从地面上挖掘出来的，由米色的石灰华作为饰面。该藏书馆可以独立运作，人们可以直接从图书馆外部进入藏书馆。

波纹玻璃立面过滤了明亮的自然光，创造出宁静的阅读氛围。铝质的天花板将漫射光引向建筑的核心地带。图书馆的外部有一个下沉天井，能够为位于地下层的员工办公室提供光线，同时也形成了一个进入图书世界之前的过渡空间。

库哈斯说："图书馆作为一种建筑类型，非常适合激进大胆的设计。显然，图书馆这种传统形式的建筑与其创造性的设计之间存在强烈的反差和矛盾，卡塔尔国家图书馆就是这样的例子。"

OMA's National Library plays a fundamental role in Doha's new academic hub, and places some of the Middle East's rarest books at the heart of its architecture.

It contains Doha's National Library, Public Library and University Library, and preserves the Heritage Collection, which consists of valuable texts and manuscripts of the Arab-Islamic civilization. It will house over a million books and space for thousands of readers over an area of 42,000m². The library plays a central role in the Education City project, a new academic campus which hosts satellite campuses from leading global universities and institutions. The master plan, designed by Arata Isozaki in 1995 and inaugurated in 2003, consists of education and research facilities, including branches of internationally acclaimed universities and the OMA-designed headquarters of the Qatar Foundation. Qatar National library is the latest expression of OMA's interest in libraries, beginning with the National Library of France competition in 1989. At that moment, the "electronics revolution" seemed "to eliminate all necessity for concentration and physical embodiment" of knowledge, questioning: would we still need them, and could they survive the digital age? In the words of the architect: "one similarity between architecture and bookmaking is that both have unbelievably long traditions but are also forced to be of the moment, constantly updating in order to survive."

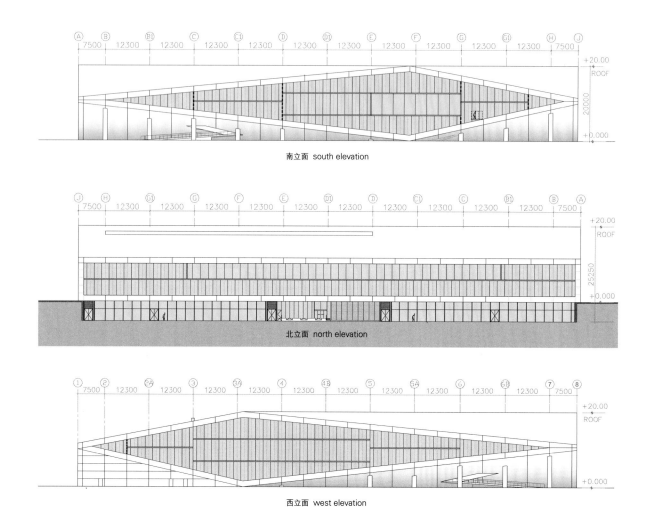

南立面 south elevation

北立面 north elevation

西立面 west elevation

The architects want to express the "vitality of the book" in a design that brings study, research, collaboration and interaction within the collection itself – among which are some of the most important and rare manuscripts in the Middle East.

The library is conceived as a single room which houses both people and books. The edges of the building are lifted from the ground creating three aisles which accommodate the book collection and, at the same time, enclose a central triangular space. According to Rem Koolhaas: "We took a plate and folded its corners up to create terraces for the books, but also to enable access in the center of the room. You emerge immediately surrounded by literally every book – all physically present, visible, and accessible, without any particular effort."

In this configuration the aisles are designed as a topography of shelving, interspersed with spaces for reading, socializing and browsing. The bookshelves are part of the building both in terms of materiality – they are the same white marble as the floors – and infrastructure – they incorporate artificial lighting, ventilation, and the book return system. A column-free bridge connects the library's main aisles, allowing for a variety of routes. It also hosts media and study rooms, reading tables, exhibition displays and a large multipurpose auditorium, enclosed by a retractable curtain designed by Amsterdam studio InsideOutside.

The heritage collection is placed at the center of the library in a six-meter-deep seemingly excavated space, clad in beige travertine. The collection can also operate autonomously, directly accessible from the outside.

The corrugated-glass facade filters the bright natural light, creating a tranquil atmosphere for reading. The diffuse light is directed into the core of the building by a reflective aluminum ceiling. Outside, a sunken patio provides light to the staff office space in the basement, and acts as transition space before entering the world of books.

"Libraries, as a typology, are so exceptionally suitable to produce radical architecture", says Koolhaas. "Apparently, there is a paradox that such a traditional form produces inventive solutions, and that is the case for the Qatar National Library."

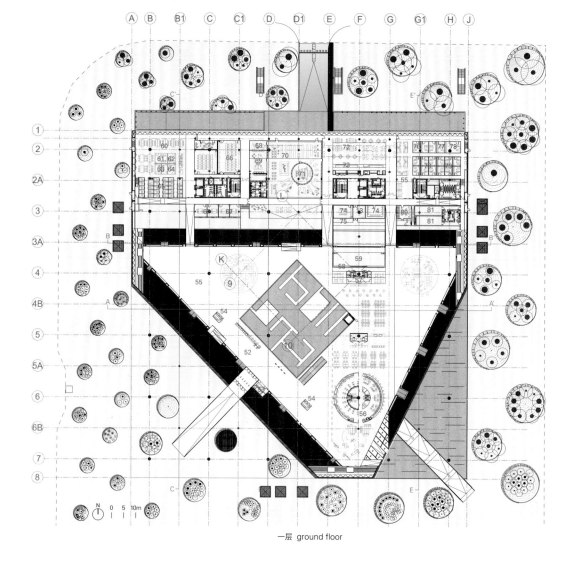

一层 ground floor

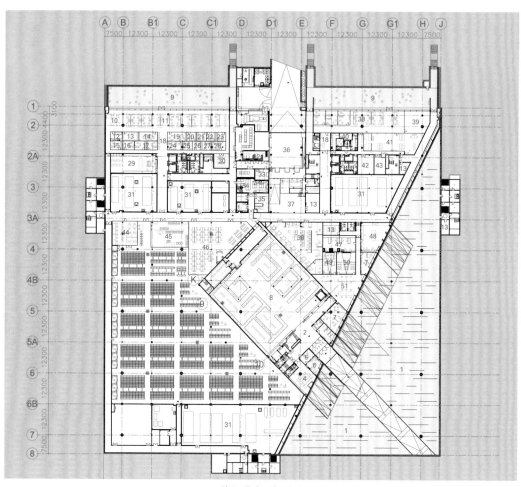

地下一层 first floor below ground

#	中文	English
1	遗产入口花园	heritage entrance garden
2	大厅与保安室	lobby & security
3	VIP卫生间	VIP restroom
4	VIP会议室	VIP meeting room
5	助理主任办公室	assistant head
6	部门负责人办公室	director head
7	专业图书管理员办公室	prof. librarian
8	遗产藏书区	heritage collection
9	天井花园	patio garden
10	休息区	breakout area
11	技术设备室与藏书办公室	tech. services & collections office
12	厨房	kitchen
13	储藏室	storage
14	资金协调与商务支持办公室	finance coordinator & business support office
15	模拟资源室	analog resources
16	数字资源室	digital resources
17	预算与支付室	budget & payment
18	接待处与休息室	reception & lounge
19	AD藏室	AD collections
20	工程书籍室	engineering
21	阿拉伯研究室	Arabic studies
22	授权专家室	licensing expert
23	艺术与人权书籍室	arts & humanities
24	采购SPC书籍室	procurement SPC
25	商务与计算书籍室	business & computing
26	社会科学书籍室	social sciences
27	文学与语言书籍室	literature & language
28	协调书籍室	coordinator
29	书籍接收区	receiving area
30	会议室	meeting room
31	设备室	plant room
32	转移室	transformer room
33	冷冻区	freeze area
34	UPS电池室	UPS battery room
35	服务器室	server room
36	卸货区	loading area
37	处理室	processing room
38	数字化室	digitization room
39	壳体空间	shell space
40	干燥储藏室	dry store
41	装订与预定空间	binding & preservation space
42	Qtel室	Qtel room
43	电信室	telecom room
44	IT办公室	IT office
45	书架整理室	sorting shelving room
46	壳体空间特殊藏书区	shell space special collection
47	技术设备室	technical services
48	保护工作区	conservation work area
49	指令发布室	instruction outreach
50	指令展示室	instruction presentation
51	研究员阅览区	researchers' reading area
52	接待处	reception
53	保安室/衣帽间	security / coat locker
54	登记处	check in / out
55	公共空间	public space
56	发行总监办公室	circulation director
57	咖啡厅	cafe
58	休息室	lounge
59	特殊活动室	special events
60	开放阅览区	open reading area
61	盲文读者室	Braille reader
62	教导主任办公室	tutoring head
63	写作室	writing room
64	写作主任室	writing head
65	学生研习间	student carrels
66	计算机室	computer classroom
67	小组学习室	group study room
68	图书馆长办公室	head librarian office
69	员工办公室	staff office
70	儿童图书馆	children library
71	多用途房间	multipurpose room
72	餐厅	restaurant
73	前厨房	front kitchen
74	媒体工作室	media studio
75	家具储藏室	furniture storage
76	书志学家办公室	bibliographer
77	R&L服务室	R&L services
78	AD公共服务区	AD. public service
79	参考书图书管理员办公室	REF. librarian
80	温室	green room
81	祷告室	prayer room
82	个人学习室	individual study room
83	女士卫生间	restroom (women)
84	男士卫生间	restroom (men)
85	礼堂	auditorium
86	展示区	display area
87	学习区	study area
88	休息空间	breakout space
89	发行办公室	circulation office
90	办公室	office
91	开放办公室	open office
92	图书馆主任办公室	library director's room
93	餐具室	pantry
94	服务休息室	service lounge
95	VIP室	VIP room
96	VIP等候室	VIP waiting
97	屋顶	roof
98	开放学习露台	open study terrace
99	专题著作室	monographs
100	开放休息露台	open lounge terrace
101	学术期刊室	scholarly journals
102	参考书目	reference books
103	视频室	video & DVD's
104	微型文本区	micro text area
105	地图室	map cases & atlas

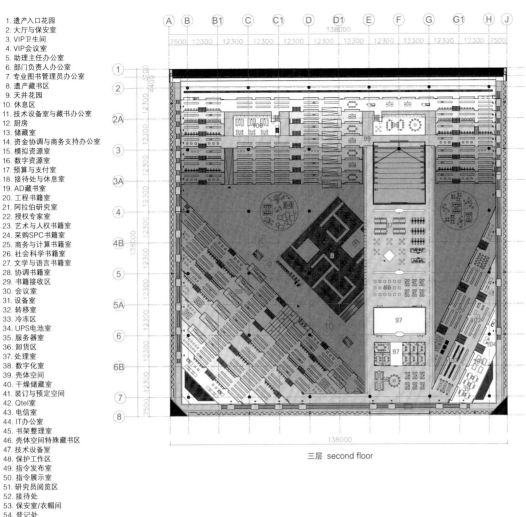

三层 second floor

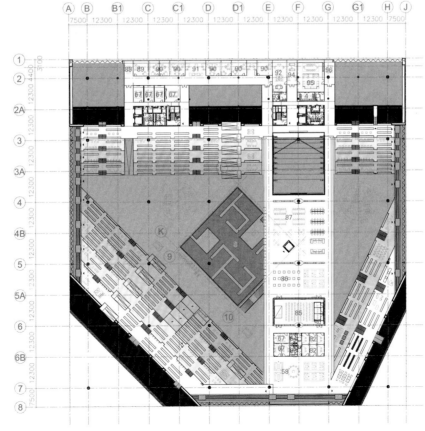

二层 first floor

A-A' 剖面图 section A-A'

B-B' 剖面图 section B-B'

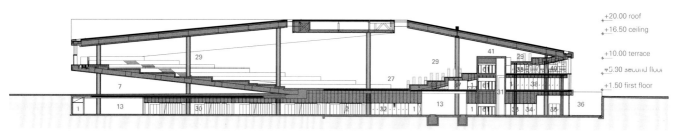

C-C' 剖面图 section C-C'

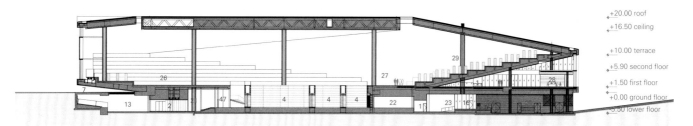

D-D' 剖面图 section D-D'

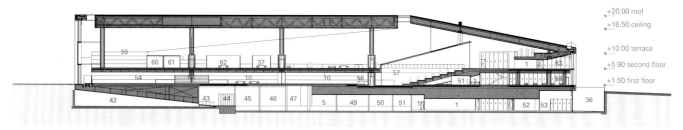

E-E' 剖面图 section E-E'

1. 走廊 2. 紧凑书架 3. 遗产展示区 4. 遗产藏书区 5. 研究员阅览区 6. 遗产入口艺术品 7. 主入口广场 8. HL设备间 9. 平台 10. 公共学习室 11. 桥梁 12. 设备区 13. 设备室 14. UPS电池室 15. 紧急照明电池室 16. 发货员室 17. 安全商店/安全UPS 18. 休息室/保安室 19. 卫生间 20. 屋顶入口通道 21. VIP卫生间 22. 壳体空间遗产藏书区 23. 处理室 24. 装货平台 25. 装货区 26. 入口服务区 27. 公共空间 28. 餐厅 29. 露台 30. 紧凑书架——主要藏书 31. 机械立管室 32. 整理室 33. 商务与计算机室 34. AD藏书室 35. 技术设备室与藏书办公室 36. 天井花园 37. 小组学习室 38. 24小时学生学习区 39. 儿童图书馆主任办公室 40. 儿童图书馆 41. 休闲区 42. 遗产入口花园 43. 遗产藏书区主入口坡道 44. 遗产藏书区主入口 45. 图书管理员助理办公室 46. 专业图书管理员办公室 47. 大厅 48. 结构书籍区 49. 指令发布室 50. 技术设备室 51. 储藏室 52. 接待处与休息室 53. 数字化室 54. 发行总监办公室 55. 部门入口 56. 咖啡厅 57. 特殊活动室 58. 展示区 59. 开放座位区 60. 女士卫生间 61. 男士卫生间 62. 教导室 63. 服务区与休息室

1. corridor 2. compact shelving 3. heritage display 4. heritage collection 5. researchers reading area 6. heritage entrance artwork 7. main entrance plaza 8. HL plant room 9. platform 10. learning commons 11. bridge 12. plant zone 13. plant room 14. UPS battery room 15. emergency lighting battery room 16. shipping clerk 17. security stores/security UPS 18. rest room/security 19. toilet 20. roof access walkway 21. VIP restroom 22. shell space heritage collections 23. processing room 24. loading dock 25. loading area 26. access service 27. public space 28. restaurant 29. terrace 30. compact shelving – main collection 31. mech riser 32. sorter room 33. business & computing 34. AD collections 35. tech, services & collections office 36. patio garden 37. group study room 38. 24 hours student study area 39. children library director 40. children library 41. recreation area 42. heritage entrance garden 43. heritage collection main entrance ramp 44. heritage collection main entrance 45. librarian assistant 46. prof. librarian 47. lobby 48. structure area 49. instruction/outreach 50. technical services 51. storage room 52. reception & lounge 53. digitization room 54. circulation dir. 55. access to department 56. cafe 57. special events 58. display area 59. open seating 60. restroom women 61. restroom men 62. L.C. tutoring room 63. service & lounge

项目名称：Qatar National Library / 地点：Doha, Qatar / 事务所：OMA / 合作伙伴负责人：Rem Koolhaas, Ellen van Loon, Iyad Alsaka / 合伙人负责人：Kunle Adeyemi
设计团队：Sebastian Appl, Laura Baird, Andrea Bertassi, Helen Billson, Benito Branco, Nils Christa, Daniel Colvard, Tom Coronato, Anita Ernodi, Clarisa Garcia-Fresco, Dina Ge, Mauricio Gonzales, Bermy H o, Vincent Kersten, Keigo Kobayashi, Dimitri Koubatis, Jang Hwan Lee, Oliver Luetjeus, Bimal Mendis, Joaquin Millan Villamuelas, Barbara Modolo, David Nam, Sebastian Nau, Rocio Paz Chavez, Francesca Portesine, Teo Quintana, Miriam Roure Parera, Peter Richardson, Silvia Sandor, Tjeerd van de Sandt, Louise Sullivan,

Anatoly Travin, Yibo Xu / 执行团队与现场团队：Vincent Kersten, Gary Owen / 分顾问：ARUP / 声学设计：DHV / 立面设计：ABT / 成本分析：David Langdon
室内设计、幕墙设计、景观设计：InsideOutside / 施工文件负责单位：CCDI / 客户：Qatar Foundation / 建筑面积：32,000m² / 总楼面积：45,000m²
设计时间：2008—2010 / 施工时间：2012.9—2017.9 / 摄影师：©Delfino Sisto Legnani and Marco Cappelletti (courtesy of the architect) - p.67, p.69, p.70~71, p.72~73, p.78, p.81 upper, p.82~83; ©Hans Werlemann (courtesy of the architect) - p.76~77, p.81 lower; ©Iwan Baan (courtesy of the architect) - p.63~63, p.64~65 left, p.68, p.80

建筑的着陆艺术

有人认为,具有特色的建筑物选址是所有建筑项目背后的主要因素之一,例如,地质和地形等有形因素是确定结构系统的关键。而诸如氛围和属性这样的无形要素,则影响建筑物的建构组成和实施。然而,当这些有形和无形的方面结合在一起时,就会出现一个关键的特征:建筑物与地面连接。

引用航空学的比喻,我们可以称之为"着陆艺术"。无数的建设者和建筑师们研究和探索了这一特点,并赋予其作品以鲜明的特色。无论是强调建筑的空灵特性还是地面属性,建筑物与地面的接触一直都是对设计的挑战。纵观历史,私人住宅作为一种建筑类型,尤其受到建筑师的青睐,并成为其尝试挑战的实验场所。这种建筑传统在许多当代项目中仍然可以看到。本节中介绍的四座住宅证明了这一点,它们展示了建筑师们如何利用构图、材料和技术的创新,完善"着陆艺术"。

The characteristics of a site are arguably some of the major factors behind any architectural project. Tangible aspects such as geology and topography are key to defining the structural system, for example. Intangible aspects such as atmosphere and identity, to name but a few, influence the project's architectural composition and materialization. There is one key feature, however, when these tangible and intangible aspects come together: the building's connection to the ground.

Using an aeronautical metaphor, we could call it "the art of landing". This particular feature has been studied and explored by countless builders and architects to attach to their creations a distinctive character. Either accentuating the building's ethereal nature or its telluric character, the contact of buildings with the ground has been a constant design challenge. Historically, private houses have been a type of building particularly for use as a locus of experimentation for architects trying to tackle this challenge. This architectural tradition can still be seen in many contemporary projects. The four houses published in this section, testify to it, showing how architects use compositional, material and technological innovations to perfect the art of landing.

奥乔克布拉塔斯住宅_OchoQuebradas House / ELEMENTAL
Bosc D'den Pep Ferrer住宅_Bosc d'en Pep Ferrer / Marià Castelló
Casa Biblioteca住宅_Casa Biblioteca / Atelier Branco Arquitetura
菲尔德霍夫住宅_Felderhof House / Pavol Mikolajcak Architekt
建筑的着陆艺术_Houses, the Art of Landing / Nelson Mota

Houses, the Art of Landing

Nelson Mota

最令人印象深刻的"着陆艺术"例子之一是令人叹为观止的14世纪建在希腊米特奥拉岩石上的拜占庭式修道院。米特奥拉修道院显示了人类的创造力在追求自然和人工制品之间几乎完美的共生方面，简直登峰造极。

着陆艺术也是建筑界争论的中心。勒·柯布西耶于1926年提出了著名的住宅设计理念——《新建筑五要素》，其中之一就是用底层架空柱将建筑抬离地面。从那时起，底层架空柱变得无处不在，成为现代建筑作品中一个显著标识和标杆，甚至在许多20世纪的不知名建筑上均可见一斑。事实上，在20世纪，建筑师和工程师通过开发和推广能够对抗万有引力的新技术方案，探索了无限的可能性将建筑与地面连接。

独户住宅经常被用来尝试做这样的实验。1929年设计的萨伏伊别墅通常被用来描述勒·柯布西耶对底层架空柱的使用，而阿德阿尔贝托·莱伯拉的马拉帕特别墅（设计于1937年）或何塞·安东尼奥·科德赫的乌加尔德别墅（设计于1953年)则被用来描述其他现代建筑师们如何试图复制米特奥拉修道院的建造者多年前完成的壮举。

One of the most impressive examples of "the art of landing" is that of the breath-taking 14th century byzantine monasteries built on the rocks of Meteora, in Greece. The Meteora monasteries show how far human ingenuity can go to pursue an almost complete symbiosis between nature and built artefact.

The art of landing has been also a central aspect in the architectural debate. One of Le Corbusier's famous *Five Points for a New Architecture*, formulated in 1926, included "the pilotis elevating the mass off the ground". Since then, the pilotis became ubiquitous, a recognizable aspect in the works that have made it into the canon of modern architecture, but also appear in many anonymous structures built during the twentieth century as well. In effect, during the last century, with the development and dissemination of new technical solutions to counter the laws of gravity, architects and engineers explored endless possibilities to connect buildings to the ground.

Single family houses were often used for such experiments. While the Villa Savoye, designed in 1929, is the usual suspect used to illustrate Le Corbusier's use of pilotis, Adalberto Libera's Casa Malaparte (designed in 1937) or Jose Antonio Coderch's Casa Ugalde (designed in 1953) illustrate how other modern architects attempted to

米特奥拉修道院，希腊
Meteora monastery, Greece

另一方面，朱利叶斯·舒尔曼为皮埃尔·凯尼格设计的第22号案例研究住宅（Stahl住宅）所拍摄的著名照片展示了建筑师和艺术家是如何对空灵产生持续的迷恋的。在舒尔曼拍摄的图片上，凯尼格设计的住宅的起居室在洛杉矶上空浮动，此图被收在了当代大多数建筑师的图片库中。最近，巴西建筑师马科斯·阿卡耶夫在1987年设计的Hélio Olga住宅中展示了如何将结构和建筑的独创性结合起来，挑战重力的基本规则，从而创造出与地面接触最少的宽敞住宅。这些著名的住宅建筑展示了建筑与地面的接触是如何在建筑设计中获得特殊地位的。本节介绍的四个案例是对"着陆艺术"扩展并探索的住宅项目代表。

Casa Biblioteca住宅（130页）是由Atelier Branco建筑事务所为巴西小镇Vinhedo设计的度假别墅，它利用场地的倾斜构造来定义空间组织和建构。屋顶被设计成一个与地面分离的水平平台，而地面则采用了遵循斜坡走向的阶梯设计。室内外空间的分隔由细长钢框架构成的玻璃立面界定，围绕着住宅三个开放的侧面设置。支撑屋顶的八根裸露混凝土柱在室内明显地显露出来，与天花板的结构网格相连。这些柱子是优雅而温暖的室内的一部分，它们的设计让人感觉这里是一个具有不同天花板高度的独立开放空间。

位于住宅上层的卫生间是唯一的封闭区域。室内放置床的区域并不是真正的"卧室"，这个区域由滑动的幕帘保持私密。住宅的主要元素图书馆——葡萄牙语为biblioteca，这个项目的名字就是由此而来的——位于夹层上，可以俯瞰厨房和用餐区，这两

replicate the feat that the builders of the Meteora monasteries accomplished so many years ago. On the other hand, the famous pictures by Julius Schulman of Pierre Koenig's Case Study House #22(Stahl House) demonstrate how architects and artists developed a constant fascination with the ethereal. Schulman's pictures of the living room of Koening's house fluctuating over Los Angeles became part of the pictorial repository of most contemporary architects. More recently, the Hélio Olga house designed by the Brazilian architect Marcos Acayaba in 1987, shows how structural and architectural ingenuity can be brought together to challenge the basic rules of gravity, creating a roomy house with minimum contact with the ground. These famous houses demonstrate how the design of the building's contact with the ground gained a special place in architectural design. The four cases featured in this section expand the portfolio of notable houses that explore the art of landing.

Casa Biblioteca (p.130), the retreat designed by Atelier Branco Arquitetura for the small Brazilian town of Vinhedo, takes advantage of the sloped conformation of the plot to define its spatial organization and composition. While the roof is designed as a horizontal platform detached from the ground, the floor is a sequence of terraces that follows the slope. The separation between exterior and interior is defined by a glass facade with a neoplastic composition of slender steel frames, surrounding the house on its three open sides. The eight exposed concrete columns that support the roof are exhibited prominently in the interior, connecting with the ceiling's structural grid. The columns are part of an elegant and warm interior, designed to be experienced as one single open space with different ceiling heights.

The block with the toilets located on the upper part of the house is the only closed area. The intimacy in the

Casa Malaparte住宅, Adalberto Libera, 意大利卡普里岛, 1938年
Casa Malaparte by Adalberto Libera, Capri, Italy, 1938

个空间位于天花板高度最高的平台上。巴西大西洋森林郁郁葱葱的绿色创造了虚拟的墙壁,界定了住宅开放的、木质的室内空间。屋顶既是一个混凝土雨篷,也是一个漂浮平台,提供了在树顶俯瞰景观的视野。Casa Biblioteca住宅由建筑师Matteo Arnone和Pep pons设计,他们二人是年轻的意大利-西班牙合资Atelier Branco建筑事务所的创始人,他们通过这个项目优雅地展示了建筑人工制品如何与场地的地形特征建立共生关系。

在过去的十年里,我们看到了极富创造力的天才建筑师们在探索实验性住宅项目时展现出来首创精神。这些项目有时也被称为"展览"。鄂尔多斯100项目就是一个例子。该项目是中国内蒙古一个城市的系列住宅项目,中国委托受赫尔佐格&德梅隆邀请的年轻建筑事务所设计。最终,这个项目失败了,只是在网络平台和建筑出版物上引起了人们的关注。位于巴塞罗那南部马塔拉纳的Solo住宅,是法国开发商克里斯蒂安·布尔代斯的一个首创项目,是这种趋势的另一个例子。受到20世纪40年代中期至60年代中期《艺术与建筑》杂志推广的案例研究住宅项目的启发,布尔达伊斯希望Solo住宅项目也能成为建筑界的一个亮点。最近,一个智利投资者发起了一个类似的首创项目——奥乔克布拉塔斯项目。奥乔克布拉塔斯项目由8名日本建筑师和8名智利建筑师设计的独户住宅组成,项目网站将项目所在场地描述为"智利北部海岸的神奇之地,距离洛斯维洛斯市4km。这个地方有大约800m的海岸线,有悬崖、海湾、岩石露头、沙丘、脆弱的短生植被,以及由阳光和海洋反射而产生的各种各样的颜色"。ELEMENTAL事务所也

area where the beds are placed – deliberately not a "bedroom" – is protected by sliding curtains. The main element of the house, the library – biblioteca, in Portuguese, from which the project gets its name – is located on the intermediate level, overlooking the kitchen and dining area, in the platform with the biggest ceiling height. The lush green of Brazil's Atlantic forest creates the virtual walls that define the open, wooden, interiors of the house. The roof performs both as a concrete canopy and a floating terrace, offering a view above the trees. With Casa Biblioteca, Matteo Arnone and Pep Pons, the founders of the young Italo-Hispanic duo Atelier Branco demonstrate with great elegance how an architectural artefact can establish a symbiotic relationship with the topographical features of the site.

Over the last decade, we have seen several initiatives interested in exploring the creative genius of architects in experimental house programs, sometimes also called "exhibitions". The case of Ordos 100, a series of housing projects for a city in Inner Mongolia is a case in point. China commissioned young architectural offices invited by Herzog & de Meuron to design this project. Eventually, the project failed and the spectacular projects came to life only in online platforms and architectural publications. The Solo Houses, located in Matarraña, south of Barcelona, an initiative of French developer Christian Bourdais, is another example of this trend. Inspired by the Case Study Houses promoted by the *Arts & Architecture* magazine from the mid-1940s to the mid-1960s, Bourdais hopes the Solo Houses program will also become a highpoint in the architecture world. Recently, a similar initiative was launched by a Chilean investor: the OchoQuebradas project. The OchoQuebradas comprises projects for single family houses designed by eight Japanese architects and eight Chilean architects for a site described on the project's website as "a magical place on Chile's northern coast, 4 km from the city of Los Vilos. The location

奥乔克布拉塔斯住宅，智利
OchoQuebradas House, Chile

Bosc D'den Pep Ferrer住宅，西班牙
Bosc d'en Pep Ferrer, Spain

被邀请参加鄂尔多斯100项目，ELEMENTAL是被邀请参加奥乔克布拉塔斯项目的智利事务所之一。ELEMENTAL的项目（90页）利用场地引人注目的景观创造了三组神秘的混凝土体量，从远处看，它们就像矗立在悬崖边缘俯瞰太平洋的几何形状岩层。水平体量容纳了住宅的主要区域，包括主卧室。这个体量最引人注目的特征之一是由超大烟囱照亮的壁炉/庭院，烟囱也被用来疏散火灾产生的烟雾。炉火给住宅的核心区域带来了温暖，平衡了墙壁和天花板上裸露混凝土的粗糙。烟囱体量经过旋转和倾斜，靠在第三个混凝土体量上，后者为客人提供了三层的小卧室和一个可以俯瞰海洋的漩涡屋顶露台。三个混凝土体量的连接方式是经过精心设计的。每个体量的比例和位置的选择惊人地精确。一方面，它显然是一个人工建筑作品；另一方面，它也可以被看作是一个神秘的图腾，就像拉帕努伊岛的摩埃石像一样，矗立在太平洋中央，距离奥乔克布拉塔斯3218.69km的地方。

与ELEMENTAL在洛斯维洛斯市设计的住宅不同，由Pavol Mikolajcak建筑事务所设计的意大利Villanders的菲尔德霍夫住宅建筑扩建结构（146页），清晰地展示了建筑物与地面的融合。绿色的草覆盖着屋后的山丘，一直延伸到屋顶，只有一条缝隙出现在屋顶，露出温暖的地下室，地下室通过宽敞的天窗照明。新扩建结构有意避免与现有的"成对农屋"冲突，这是艾萨克山谷斜坡上的典型建筑。在尊重现有的相邻本地建筑的基础上，建筑师精心设计了独特的室内氛围，可以将白云石山脉的顶峰尽览无余。该项

has approximately 800 meters of coastline, with cliffs, bays, rocky outcrops, sand dunes, fragile and ephemeral vegetation, and an infinite variety of colors produced by the sunlight and the reflections from the sea.'" ELEMENTAL, which had also been invited to participate in the Ordos 100 program, was one of the Chilean offices invited to participate in OchoQuebradas. ELEMENTAL's project (p.90). takes advantage of the site's dramatic landscape to create a set of three uncanny concrete volumes that, seen for afar, resemble geometric rock formations standing at the verge of the cliff overlooking the Pacific Ocean. The horizontal volume accommodates the main areas of the house, including the master bedroom. One of the most striking features of this volume is the fireplace/courtyard lit by an oversized chimney, which is also used to evacuate the smoke produced by the fire. The fire brings the warmth to the core of the house, balancing the roughness of the exposed concrete used in the walls and ceilings. The volume of the chimney is rotated and tilted against the third concrete block, which accommodates small bedrooms for the guests, staked in three floors, and a roof terrace with a whirlpool overlooking the Ocean. The articulation of the three concrete volumes is carefully crafted. The precision used to find the right proportion and location for each volume is striking. On one hand, it is clearly an architectural artefact. On the other hand, it can also be seen as a mysterious totem, like the Moai of Rapa Nui, that stand 3218.69 km in front of OchoQuebradas in the middle of the Pacific Ocean.

As opposed to ELEMENTAL's house in Los Vilos, the extension of a residential building at the farmstead Felderhof House (p.146) in Villanders, Italy, designed by Pavol Mikolajcak Architekt, shows a clear attempt to merge with the ground where it was built. The green grass that covers the hills behind the house continues to its roof and it is only broken by a slit that reveals a warm underground interior, lit by generous skylights. The new extension deliberately avoids a confrontation with the existing "pair farmstead" typical of the slopes of the Eisack

菲尔德霍夫住宅，意大利
Felderhof House, Italy

目不是寻求与景观的有机结合，而是探索与地面的互动，将地面作为媒介来激发新旧建筑、黑暗与明亮空间、硬与软表面之间有趣的过渡。

　　福门特拉岛的Bosc d'en Pep Ferrer住宅（110页）由西班牙建筑师Marià Castelló设计，利用场地的地质特征，来建造能够探索地面构造和地质构造之间二元性的项目。该住宅位于福门特拉岛南海岸的米格约恩海滩附近，较低的一层是一个洞穴，让人想起马略卡岛海岸的母马采石场。住宅的承重结构将现有的地质元素和两面平行支撑墙的混凝土构件结合起来。在这一层之上，有三个独立的原始木质体量，由玻璃走廊连接。这些体量由交叉层压木板建成，容纳了住宅的起居和睡眠空间，并有助于界定低层的一系列天井。不同的天井提供了不同的空间体验，在建造过程中发现的洞穴创造了不同寻常的场景，强化了这种体验。住宅露出地面的部分与邻近的皮德斯卡塔拉塔楼（建于1763年）的外形构成呼应。Marià Castelló设计的住宅同时也是对地面的明显改造，体现了虚实的融合，与福门特拉的地中海景观形成对比。

　　这四个项目既展示了"着陆艺术"在诠释房屋与地面连接时的不同处理方式，又展示了仔细解读场地拓扑特征及其现象学意义的重要性。

valley. Despite its respect to the existing neighboring vernacular structures, the architect created a distinct domestic atmosphere crafted to explore the spectacular views to the peaks of the Dolomites. Rather than seeking an organic integration with the landscape, this project explores the interaction with the ground as a device to stimulate playful transitions between old and new constructions, dark and light spaces, hard and soft surfaces.
Bosc d'en Pep Ferrer, the house in Formentera Island (p.110), designed by Spanish architect Marià Castelló, takes advantage of the geological features of the site to build up a project that explores the duality between the telluric and the tectonic. Built on a plot located next to the beach of Migjorn, on the south coast of the island of Formentera, the lower level of this house is shaped as a cavity reminiscent of the Marès quarries on the Majorcan Coast. The load-bearing structure of the house combines the existing geological elements with concrete elements defining two parallel supporting walls. Above this level, there are three detached pristine wooden volumes connected by glazed galleries. These volumes, built with cross-laminated wood panels, contain the house's living and sleeping areas, and contribute to define a sequence of patios in the lower level. Each of these patios provides a different spatial experience, intensified by the unusual scenario created by the caves discovered during construction. The volumes that stand out, above the underground level, establish a dialogue with the silhouette of the neighboring Pi des Català Tower, built in 1763. Marià Castelló's house is at the same time a conspicuous transformation of the ground and a sensible composition of solids and voids set against Formentera's Mediterranean landscape.
These four projects show different approaches to the "art of landing" in housing on the ground. They demonstrate the importance of "reading" carefully the topological characteristics of the site, but also its phenomenological meaning.

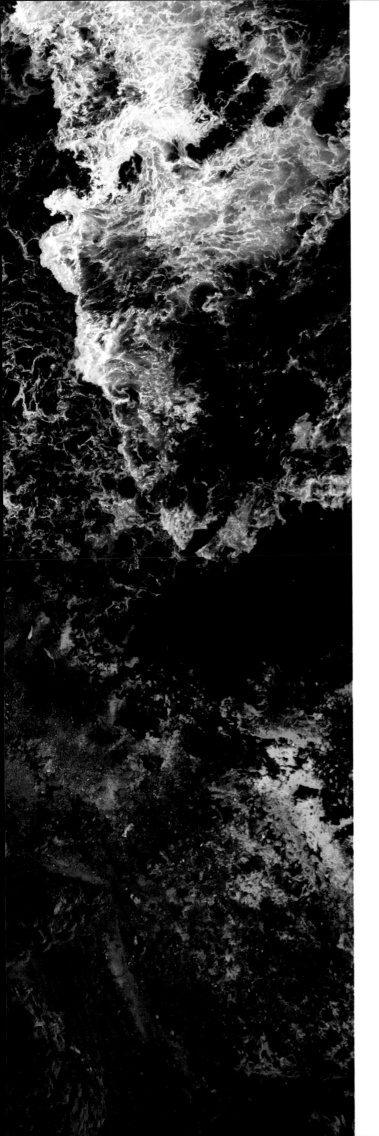

奥乔克布拉塔斯住宅
OchoQuebradas House

ELEMENTAL

智利奥乔克布拉塔斯住宅的元素设计
An elemental design for OchoQuebradas House, Chile

ELEMENTAL事务所在坐落于智利太平洋海岸的奥乔克布拉塔斯住宅的设计中探索了原始的本质。

奥乔克布拉塔斯（八个峡谷）是智利太平洋海岸的一个私人开发项目，位于圣地亚哥以北250km处。它汇集了8名日本建筑师和8名智利建筑师的设计，每位建筑师都在该地区建造了一座周末别墅。目前还没有个人客户。开发商定义了建筑面积（250m²）、功能空间（包括四间卧室、客厅和餐厅、厨房、浴室和一间酒窖。）、总体预算（50万美元）。每位建筑师都拥有完全的自由根据以上要求进行设计。

ELEMENTAL事务所将项目场地和住宅将作为周末别墅住宅使用这一事实视为探索某种原始的机会。那里的地理环境非常恶劣，只有一组坚固粗糙的构件似乎才合适。因为海浪到达陆地时猛烈的撞击，这里的海洋是白色的。而周末别墅终究该是一个静修之地，在这里，人们可以回归到更本原的生活状态。建筑师将"桌子另一边的缺席"（即没有人）作为辩解，设计抛却家庭生活的传统，探索了生活中不可或缺的维度。他们选择性地回顾过去，不是为了逃避怀旧，而是为了自然地过滤陈词滥调。在这个时代，对新奇的渴望威胁着建筑物的时效，使建筑几近废弃的边缘，而建筑师力图寻找的是永恒的设计。ELEMENTAL事务所设计了三个体量：一个是水平体量，略微悬挑在悬崖的顶部，一对夫妇不需要打开房子的其余部分就可以自给自足；第二个是垂直体量，包含客户需要的其他房间；设在屋顶的露台减少了项目占地面积，且增加了客户饱览浩瀚海洋景观的视野；在这两者之间有一个略微倾斜的中空体量，里面有一堆火，它不是烟囱（烟囱已经成为文明的产物），而是火种（人类最具革命性和最原始的成就之一）。这些结构的五个墙面由混凝土浇筑而成，第六面由与混凝土模板相同的木材构成。这样设计的目的是让这些结构能够像石材那样慢慢老化，呈现出场地所特有的某种蛮荒气质，却又不失温暖柔和，让人能够享受普通的自然和生活。

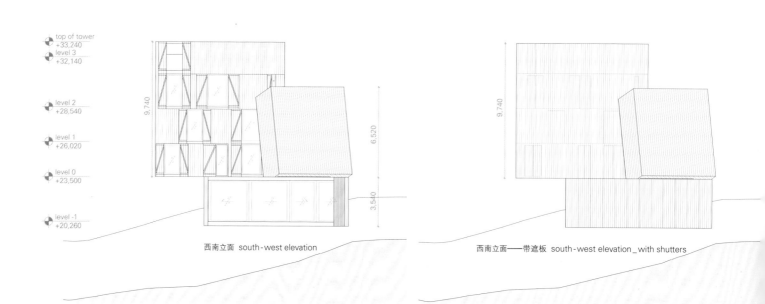

西南立面 south-west elevation　　　西南立面——带遮板 south-west elevation_with shutters

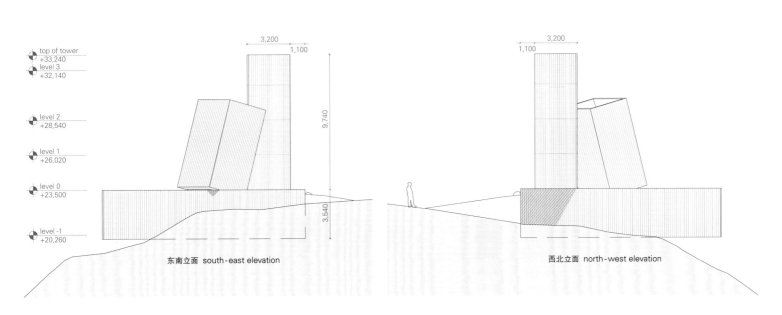

东南立面 south-east elevation　　　西北立面 north-west elevation

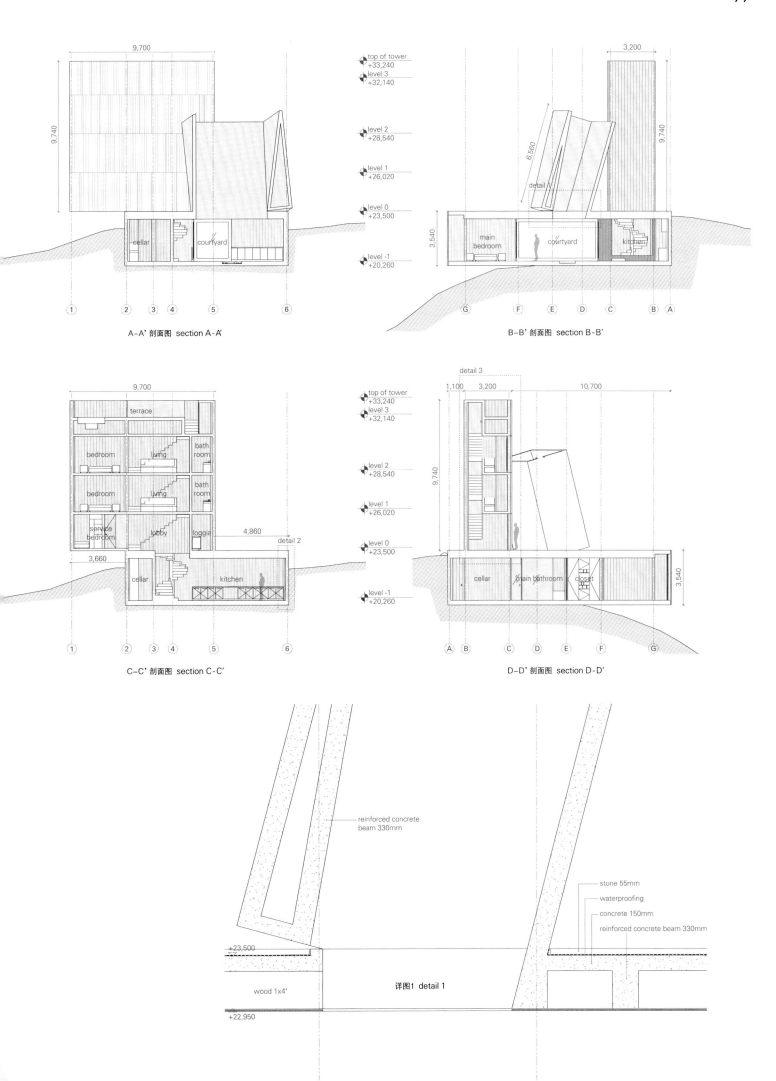

项目名称：Ochoquebradas House
地点：Los Vilos, Chile
事务所：ELEMENTAL
项目团队：Alejandro Aravena, Victor Oddó, Suyin Chia
合作者：Alexander Frehse, José Esparza, Carlos Portillo, Isaías Moreno, Clémence Pybaro
结构工程师：Luis Soler P. & Associates
总承包商：Eric Meinardus–MJA Arquitectura y Construcción Ltda.
机电管道工程师：Geocav Ltda.
卫浴工程顾问：Patricio Moya
电气工程顾问：Julio Rojas
供暖顾问：Ingesec EIRL
窗户供应商：Raumdesign, Schüco partners Chile
乙酰化木材供应商：Leaf Panels Spa
用地面积：5,355m² / 建筑面积：289m²
结构：reinforced concrete
材料：concrete, wood, glass
设计时间：2013—2014 / 施工时间：2017—2018
摄影师：©Christobal Palma

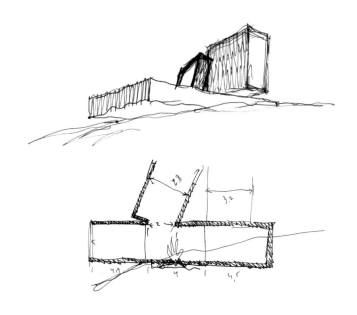

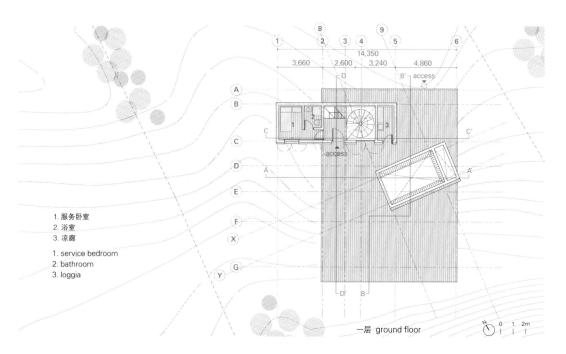

1. 服务卧室
2. 浴室
3. 凉廊

1. service bedroom
2. bathroom
3. loggia

一层 ground floor

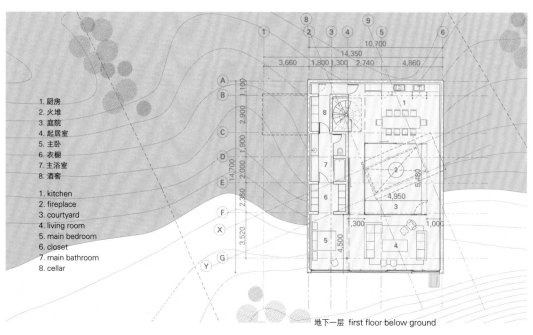

1. 厨房
2. 火堆
3. 庭院
4. 起居室
5. 主卧
6. 衣橱
7. 主浴室
8. 酒窖

1. kitchen
2. fireplace
3. courtyard
4. living room
5. main bedroom
6. closet
7. main bathroom
8. cellar

地下一层 first floor below ground

ELEMENTAL explore the nature of the primitive in their designs for OchoQuebradas House on Chile's Pacific Coast.

OchoQuebradas (Eight Ravines) is a private development on the Pacific coast of Chile, 250km north of Santiago. It brings together eight Japanese architects and eight Chilean architects, each building a weekend house in the district. As yet there are no individual clients; the developer has defined a built area (250m²), a program (four bedrooms, living and dining area, kitchen, bathrooms and a wine cellar) and an overall budget ($0.5 million) that each architect may respond to complete freedom.

ELEMENTAL viewed the site and the fact that the dwellings are intended to be weekend houses, as an opportunity to explore a certain primitiveness. The geography was sufficiently brutal that only a strong, rugged, set of elements seemed appropriate. The Ocean here is white due to the violence of the waves crashing as they reach the land. Yet a weekend house is ultimately a place of retreat where people allow themselves to go back to a more essential style of living. The architects used the "void on the other side of the table" (i.e. the absence of a client) as an alibi to

eliminate the conventions of domestic living, exploring instead more irreducible dimensions of life. They chose to look backwards, towards not as a nostalgic escape but as a natural filter against time-worn clichés. In an era where the hunger for novelty is threatening architecture to the point of obsolescence, the architects looked for timelessness. ELEMENTAL designed three volumes: a horizontal one, slightly cantilevering on top of the cliff and self-sufficient for a couple to use without the need to open the rest of the house. Then a vertical volume, containing other rooms required by the clients. A terrace on top reduces the footprint of the site and expands the horizon towards the vastness of the ocean. In between these two sits a slightly leaning and hollowed volume containing a fire, not a chimney (which is already something civilized), but a fire (one of the most revolutionary and primitive achievements of man).

Five sides of the pieces are made of poured concrete; the sixth is made out of the same wood which was used as a formwork for the concrete. The intention is that these pieces can age in a similar way that stone ages, acquiring some of the brutality of the place but still gentle enough for people to enjoy nature and life in general.

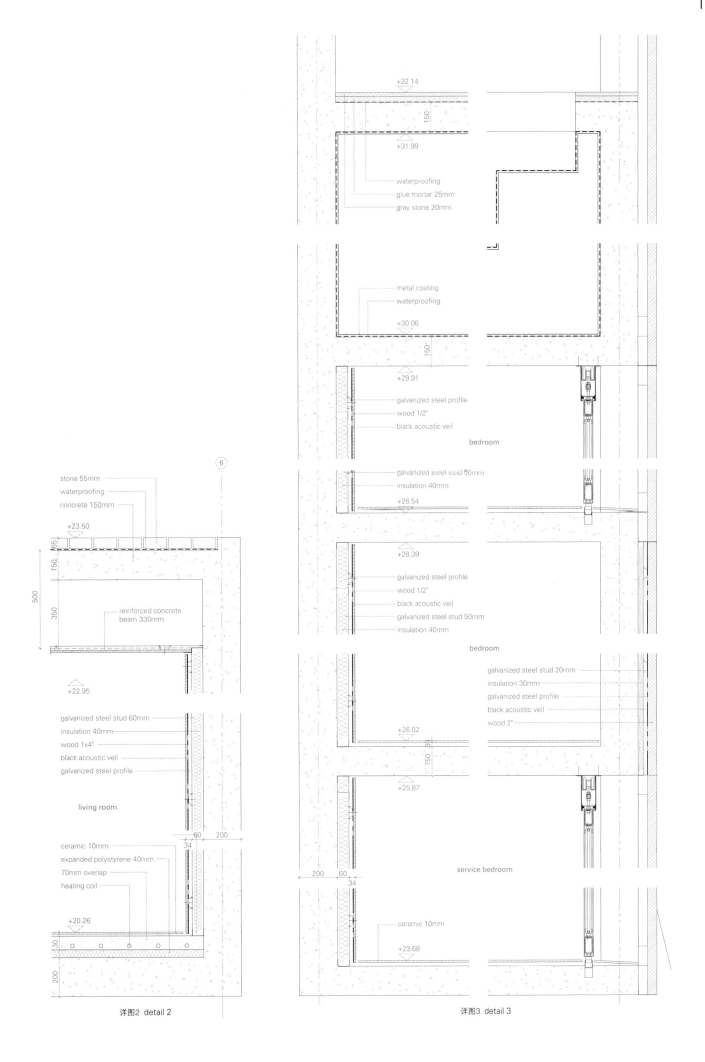

Bosc d'en Pep Ferrer 住宅
Bosc d'en Pep Ferrer

Marià Castelló

整体、巨石和立体；巴利阿里群岛梦幻景观中的住宅
Monolithic, megalithic and stereotomic; house in the dreamy landscape of the Balearic Islands

Bosc d'en Pep Ferrer是位于Migjorn海滩附近的一块土地的传统名称，它位于巴利阿里群岛的福门特拉南海岸。在这片梦幻般的风景中，地平线上唯一的干扰就是建造于1763年的皮德斯卡塔拉塔楼。

该项目侧重于二元性：地形和地质、重和轻、大地和空气、手工和技术、压缩和牵引、努力和阻力。

地面上的岩石表面经过雕刻，就像一个雕塑，提供了一个能让人想起"母马"采石场的洞穴，这是一个从石头演化而来的空间。

建筑是整体的、巨石状的、立体的。它提供了一座家庭住宅，一座根据地形和环境而设计的住宅。

住宅被分成三个轻质模块，用干施工技术建造，住宅下面有一个空腔，通过掏空下层材料形成。纵向布局的建筑位于一个洞穴之上，洞穴是设计的一个组成部分。天井、连接通道和横向的景观给人以凌空感并令人叫绝。这个结构很容易理解：它由三层组成，越往上建筑精度越高。最下面一层明显没有外墙，因为它是由岩石表面构成的。上面是混凝土结构，包括一层和支撑结构，用来支撑上面的楼层。在有双层支撑的上层，一个元素（交叉层压木板）同时发挥了多种功能：稳定结构、封闭空间和室内装饰。

高规格的材料以及它们搭配的方式，在项目中扮演着重要的角色，特别是设计还强调"生物构造"和天然材料与本土材料。这些材料包括雕刻的岩石、挖掘工地产生的碎石卡普里石灰石、松树和杉木、再生棉板、白色Macael大理石、高渗透性硅酸盐涂料等。

这些材料有助于创造吸湿性围护结构和房间，它们可以渗透水分，如水和蒸汽，还能创造一个愉快、健康、节能的室内环境，这样创造出的被动生物气候系统在这里的气候条件下被证明是有效的。水的自给自足是通过一个大容量的雨水池来实现的。此次建筑干预措施的目的是用天然材料来创建一个重要的庇护所，以此来表达对环境的尊重。建筑师呼吁在设计中体现当地建筑的传统价值，如水和能源的自给自足、景观的和谐性和资源的优化，以此来传递福门特拉岛文化的特殊内在价值。

Bosc d'en Pep Ferrer is the traditional name for the plot of land located next to Migjorn beach, on the south coast of Formentera in the Balearic Islands. In this dreamy landscape, the only interruption on the horizon is the silhouette of the Pi des Català Tower, built in 1763.

The project focuses on dualities: telluric and tectonic, heavy and light, earth and air, handcrafted and technological, compression and traction, effort and resistance.

The rocky surface of the terrain has been carved, as if it were a sculpture, offering a cavity reminiscent of the "marès" stone quarries: a space materializes from the stone.

The architecture is monolithic, megalithic and stereotomic. It offers a family house, sensitive to the topography and environment.

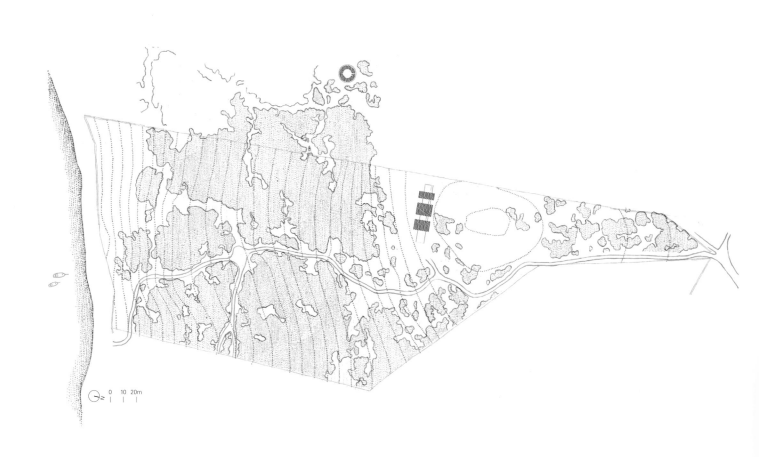

The house is divided into three lightweight modules built with dry construction techniques above a cavity made by subtraction of material from the lower floor. This longitudinal placement sits atop a cave, an integral part of the design. Patios, connecting walkways and transverse views create a place defined by time and characterized by surprises. The structure is easily comprehensible: it is made up of three strata, with ascending levels of architectural precision. The lower floor expresses the obvious absence of containment walls, as it materializes from the rock surface. Above it is the concrete structure which constitutes the ground floor and the support structure for the level above it. On the doubly-supported upper floor, single element (cross-laminated wood panels) clusters several functions: structure, closure, and interior finishing.

A high specification of materials, and the ways they have been paired, plays an important role in the project, in particular the emphasis on "bioconstruction" and natural, local

materials. These include sculpted rock, crushed gravel generated from the excavation of the site, Capri limestone, pine and fir wood, recycled cotton panels, white Macael marble, high permeability silicate paints, etc.

These materials have contributed to the creation of hygroscopic enclosures, rooms which are permeable to moisture, such as water and steam, and which create a pleasant, healthier and energy efficient indoor environment: passive bioclimatic systems, which have proved effective in this climate. Water self-sufficiency has been achieved thanks to a large-volume rainwater cistern that reuses rainwater. This architectural intervention seeks to create a vital shelter, built with natural materials, which is respectful to the environment. The architects called on traditional values of local architecture, such as water and energy self-sufficiency, harmony with the landscape and the optimization of resources, in order to transmit some of the intrinsic values of the culture of the island of Formentera.

屋顶 roof

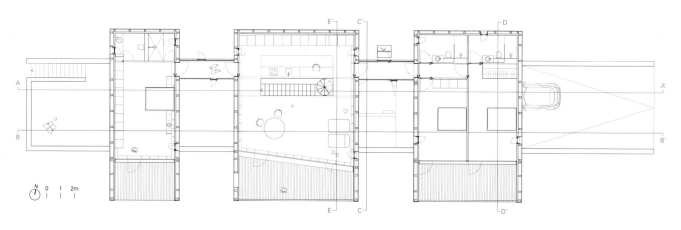

一层 ground floor

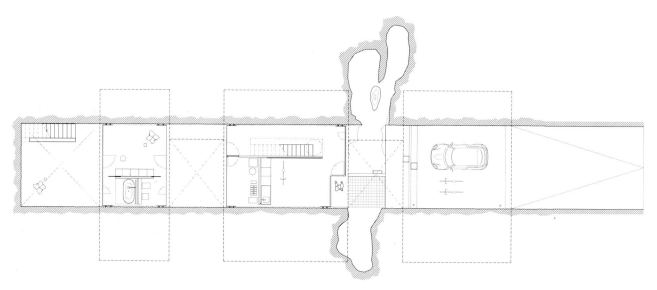

地下一层+天然洞穴 first floor below ground + natural cave

项目名称：Bosc d'en Pep Ferrer / 地点：Migjorn beach, Formentera, Spain / 事务所：Marià Castelló / 建筑工程师：Agustí Yern Ribas
结构工程师：Miguel Rodríguez Nevado, Ferran Juan / 设备工程师：Javier Colomar Riera / 合作者：Marga Ferrer, Natàlia Castellà, Lorena Ruzafa, Elena Vinyarskaya / 建筑商：Motas Proyectos e Interiorismo S.L., Luis Tulcanazo Castro, Antonio Serra Requena, Foreva S.L. / 分包商：iCarp Valencia S.L., Velima System S.L., Astiglass S.L., Singularglass S.L. / 总楼面面积：243,59m² on ground floor + 71,73m² in basement floor
设计时间：2007—2014 / 施工时间：2014.11.4—2017.3.15 / 摄影师：courtesy of the architect

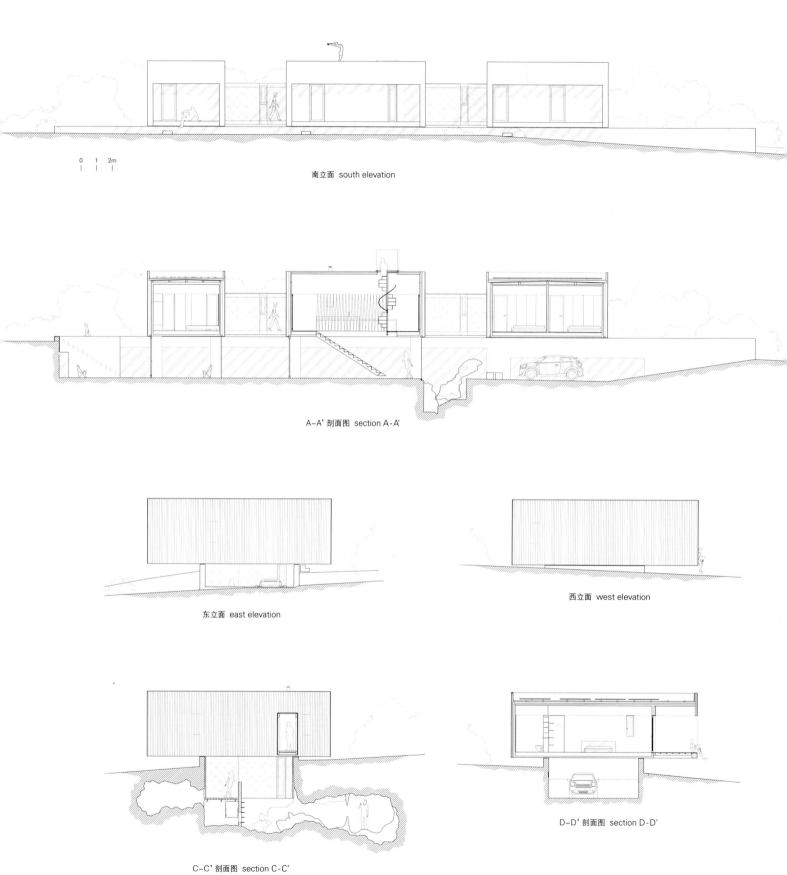

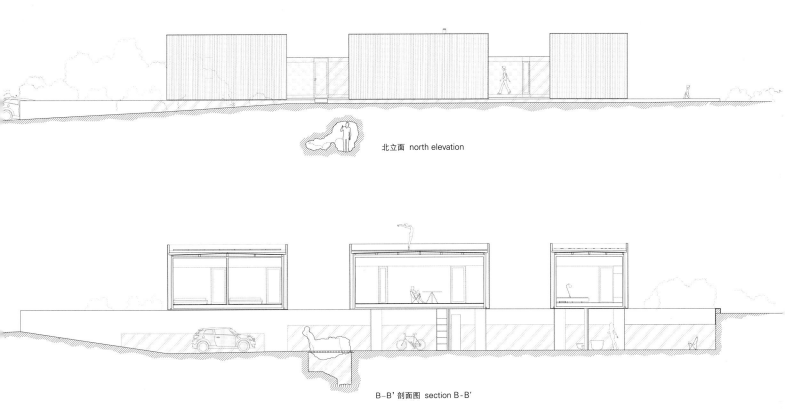

北立面 north elevation

B-B' 剖面图 section B-B'

1. main structure of CLT of fir wood and pine (only on ground floor), visible from the inside
2. ventilated facade of planks, battens and pegs of larch pine wood
3. breathable thermal insulation made of recycled cotton plates (10cm), protected with a breathable and waterproof sheet
4. WBP plywood panels painted white with silicate paint
5. battens for pavement and photovoltaic panels support made of larch pine wood
6. EPDM waterproof sheet
7. OSB class 3 boards as support for the waterproof sheet
8. breathable thermal insulation made of recycled cotton plates (10cm)

建筑构造立面详图 building construction facade detail

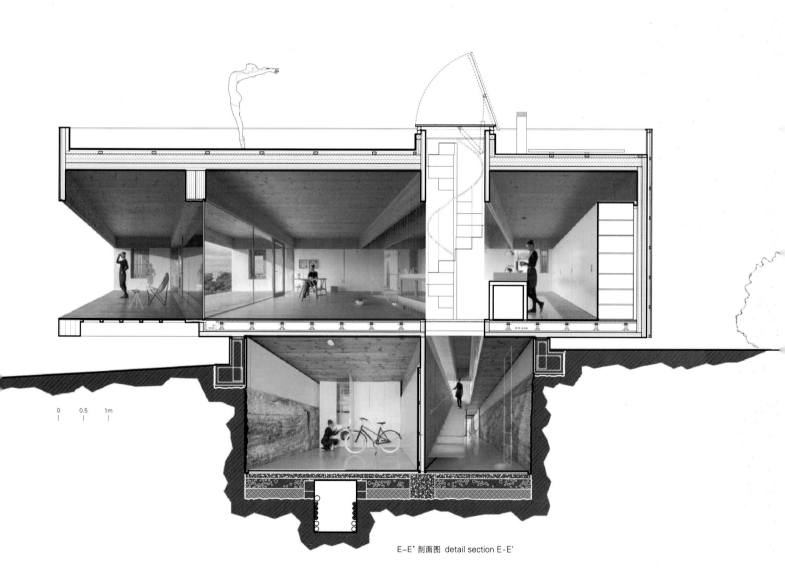

E-E' 剖面图 detail section E-E'

Casa Biblioteca 住宅
Casa Biblioteca
Atelier Branco Arquitetura

Casa Biblioteca 以20世纪50年代的"圣保罗"主题为基础，创造了一个休闲、沉思的家
Casa Biblioteca draws on 1950s "Paulistana" motifs to create a leisurely and contemplative home

2014年夏天，有人找到Atelier Branco建筑事务所的Matteo Arnone和Pep Pons，希望他们在巴西的Vinhedo设计一个休闲度假场所；设计纲要要求有一个阅读的地方——沉浸在场地生机勃勃的植被中——一个思考的地方，就在无边无际的亚热带天空下。

客户年轻时是一位反对巴西军事独裁的左翼活动人士，现在是著名的政治思想史学者，他将这座住宅视为一个临时的天堂，位于圣保罗和他的工作地坎皮纳斯大学之间。因此，这座房子既不是永久性的住宅，也不是传统的度假屋，而是一个沉思的地方，偶尔也是一个远离巴西大都市喧嚣的工作场所。

由于客户的需求和周边的环境都不寻常，因此Casa Biblioteca住宅是一个特殊的项目。它坐落在一块朝北的地形陡峭的地块内，位于Vinhedo茂密的mata atlantica森林的一块空地上，这片森林覆盖了巴西大西洋沿岸地区。由于地形属性的原因，设计遵循了明显的"剖面"原理，因此项目的空间和功能配置几乎完全由两条等高线之间的关系来表达。这两条线中的第一条是地面的线条，倾斜的轮廓被处理成一系列宽敞的水平阶梯；第二部分是屋顶的线条，屋顶稍稍悬浮在最上面的挡土墙上，在经过人工处理的地形和天空之间创造了一个锐利的水平基准。

混凝土屋顶对整个项目的连接起着至关重要的作用，表现了外部和内部的外观。它是一块细长的、15cm厚的矩形板，由八根柱子支撑。当人从主干道向住宅走近时，首先看到的是屋顶板的上面，那里是一个巨大的观景台，四周被树冠的叶子包围。平台上铺设了加拉皮拉木板，木板与平台的对称线对角相接，平台上没有设置女儿墙，取而代之的是由1m宽的水池环绕，水池界定了一个矩形的中心岛，从那里可以欣赏到周围的景色。

屋顶周边有缺口的地方设置了一部楼梯，通向住宅的中央空间。这是一个未分割的、全玻璃的长方形房间，容纳了全部的室内活动。从最私密的区域往下走便来到更开放的区域，房屋高度也逐渐增加。由于现有斜坡不平坦，因此使剖面呈现出独特锯齿状轮廓的挡土墙的高度非常独特，但其间距相等，从而形成了一个三层结构体系。

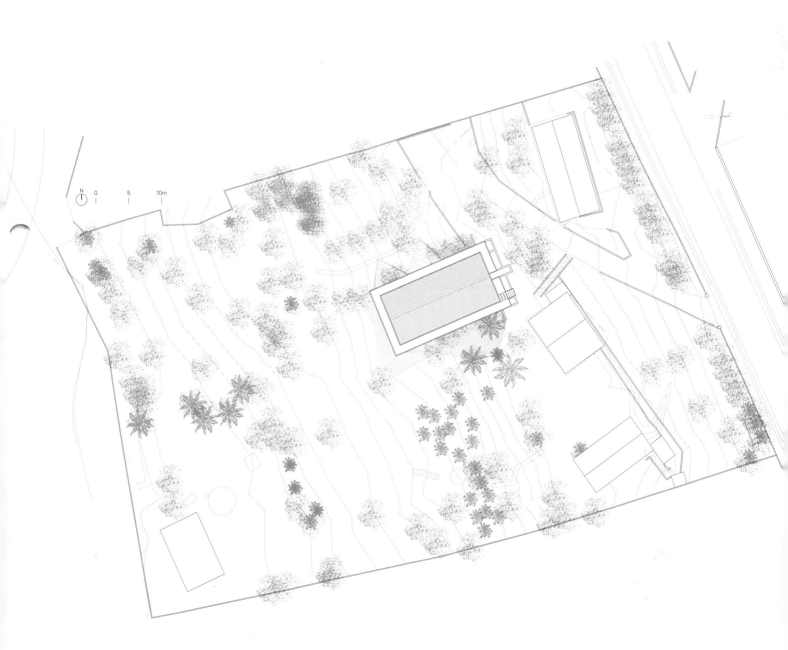

虽然深度相同,但三个台阶均是从地面到天花板的高度,能提供足够的活动空间,满足保护隐私的需要,同时兼顾充足的采光。卧室位于项目最上层的平台上,这一层能俯瞰住宅的中央平台,中间这层是客户的工作室,此区域与景观直接相连,两边各有两扇设置在中央的玻璃门。最后,离入口最远的平台是可以俯瞰周围绿地的起居和用餐区,这是住宅里最明亮、最暴露的地方。

类似的处理方法也适用于核心设备区和储物设施的位置设置,这些设施或被嵌入挡土墙内,或被精确地靠在挡土墙上,例如厨房构件、书架和衣柜,因此,空间不受与设计无关的元素的束缚。

材料的选择和处理使住宅拥有一种原始的庄严感。裸露的结构由钢筋混凝土建成。地板和上层露台由加拉皮拉木板铺设而成,铺设方向垂直于项目的纵轴。整个结构几乎完全包裹在单层玻璃立面中,其坚固的外形和设计主题遵循了20世纪50年代标志性的"保罗-利斯塔纳"传统。

Casa Biblioteca住宅以其精细的细木工和准空灵的开放性,融合了玻璃屋设计的丰富传统。自20世纪初以来,玻璃屋设计一直被用于建筑实验中。

In summer 2014, Matteo Arnone and Pep Pons of Atelier Branco were approached to design a leisurely retreat in Vinhedo, Brazil; the brief called for a place to read – immersed in the site's vibrant vegetation – and a place to think, under boundless subtropical skies.

The client, a left-block activist against Brazil's military dictatorship in his youth and now renowned scholar of the history of political thought, had viewed the house as a

temporary haven between São Paulo and his workplace at Campinas University. The house is therefore not a permanent residence nor a conventional holiday home, but a place of contemplation, occasionally of work, away from the bustle of metropolitan Brazil.

Casa Biblioteca is an idiosyncratic project, due to the eccentricity of its client and the context. It is set on steep north-facing terrain within a clearing of Vinhedo's dense "mata atlantica" – the forest which extends over Brazil's Atlantic littoral region. Due to topographical attributes, the design follows a distinctly "sectional" rationale such that the spatial and functional disposition of the project is almost entirely articulated in the relation between two contour lines.

The first of these two lines is that of the ground, where the sloping profile has been manipulated to form a series of spacious horizontal terraces; the second consists of the line of the roof, which slightly hovers over the uppermost retention wall to create a sharp, horizontal datum between the domesticated topography and the sky.

The concrete roof plays a crucial role in the articulation of the program, in the characterization of its external and internal appearance. It is a slender, 15cm-thin rectangular slab

supported by eight pillars. When approaching the house from the main road, it is the upper face of this element that presents itself to the viewer, offering access to a monumental viewing platform within the foliage of surrounding tree canopies. In place of a parapet, the deck – lined with Garapeira wood boards which meet diagonally against the deck's symmetry line – is circled by a meter-wide water bed which defines a rectangular central island from which to contemplate the view.

A dentil in the perimeter of the roof allows for a staircase leading into the house's core space. This is an undivided, fully glazed, rectangular room hosting the entire domestic program; the height gradually increases as one descends from the most intimate to the more exposed areas of the home. The retention walls which give its section the distinguished jagged profile are unique in height due to the uneven slope of the existing terrain, but spaced equally to create a tripartite structural system.

Although equal in depth, the three terraces gain unique floor-to-ceiling heights providing the activities which take place on them with degrees of privacy and natural lighting conditions. Bedrooms are located onto the project's uppermost terrace; this level overlooks the house's central platform which hosts the client's studio, the area of the house most directly connected to the landscape with two centrally-placed glass doors at either side. Lastly, the terrace furthest from the entrance is a living and dining area overlooking the surrounding greenery; this is the brightest and most exposed area of the house.

A similarly methodical approach informs the location of the core services and storage facilities which are either carved into the project's retention walls or fitted accurately against them – as with the kitchen elements, bookcases and wardrobes. As such, the space remains untethered from elements foreign to its "disegno".

The selection and treatment of materials give the house a rudimentary grandeur. The bare structure is reinforced concrete. The floors and upper terrace are lined with Garapeira wood boards placed perpendicular to the project's longitudinal axis. And the whole structure is almost entirely wrapped within a single-glazed facade of which the iron profiling and design motifs follow the iconic 1950s "paulistana" tradition.

With its fine joinery and quasi-ethereal openness, the Casa Biblioteca joins a rich tradition of glass houses which since the dawn of the 20th century have lent themselves to architectural experimentation.

项目名称：Casa Biblioteca
地点：Vinhedo, São Paulo, Brazil
事务所：Atelier Branco Arquitetura
项目负责人：Matteo Arnone, Pep Pons
项目团队：Andreas Schneller, Cristina Plana, Marta Pla, Martina Salvaneschi
结构工程师：John Biscuola – Biscuola Engieneering
基础工程师：Sergio Ludemann – Ludemann Engieneering
系统工程师：João Claudinei Alves – JCF projetos e construções
模板与木工：Edivaldo Lourenço and Eduardo Lourenço
功能：private residence to accommodate the need for a place to read and the need of a place to think
结构系统：In-situ cast concrete, all cast within a single working day
主要材料：cast concrete, steel, glass, and garapeira wood
用地面积：5,500m²
建筑面积：150m²
总楼面面积：200m²
建筑造价：R$ 400.000,00 (U$ 125.000,00)
设计时间：2013.3—2014.2
施工时间：2014.6—2015.11
摄影师：
©Gleeson Paulino (courtesy of the architect) - p.133, p.134, p.135, p.137^{upper}
©Jaqueline Lessa (courtesy of the architect) - p.130~131, p.136
©Ricardo Bassetti (courtesy of the architect) - p.137^{lower}, p.138~139, p.142~143, p.144~145

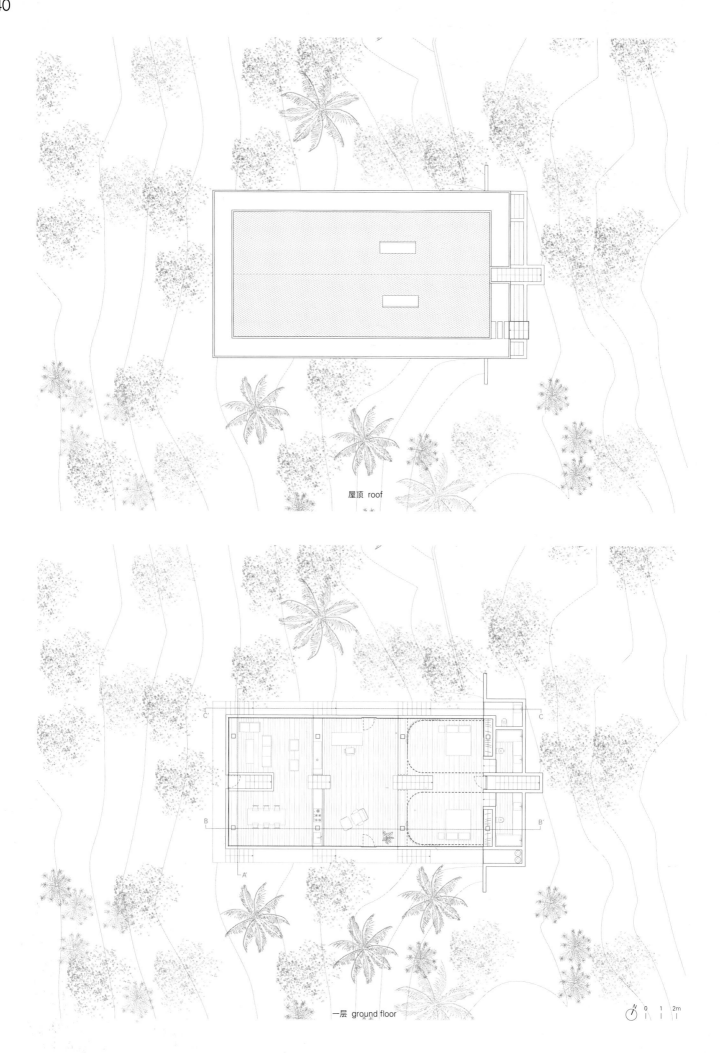

屋顶 roof

一层 ground floor

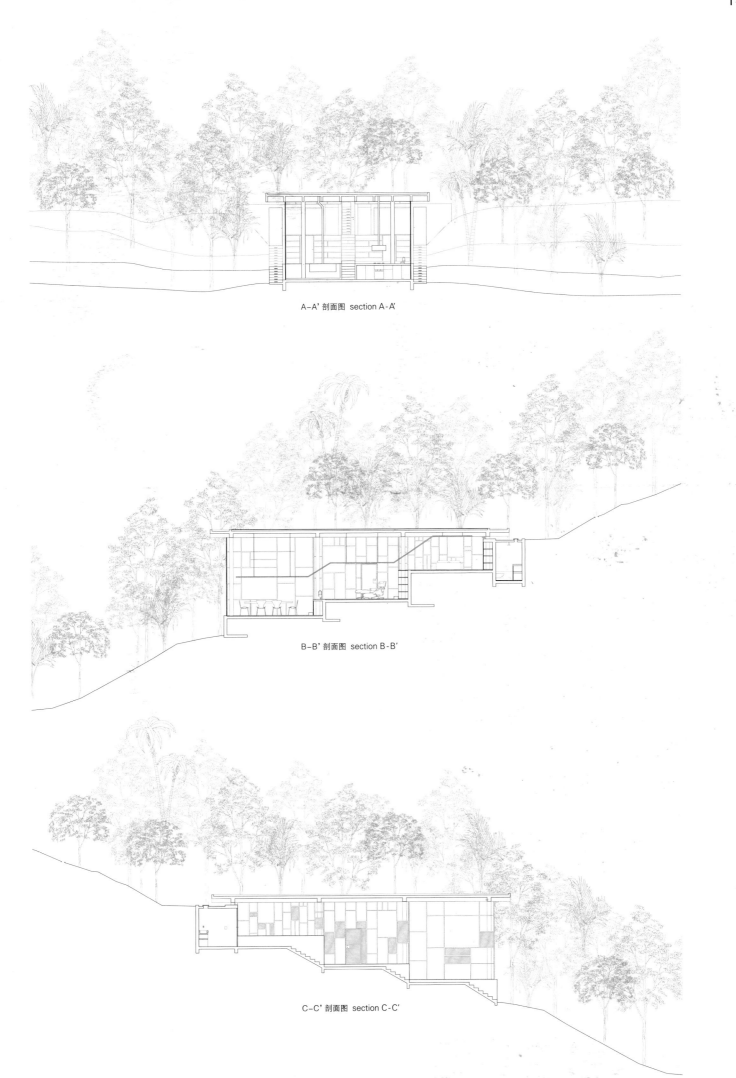

A-A' 剖面图 section A-A'

B-B' 剖面图 section B-B'

C-C' 剖面图 section C-C'

菲尔德霍夫住宅
Felderhof House
Pavol Mikolajcak Architekt

菲尔德霍夫——意大利北部的一个传统农场
Felderhof – extending a traditional farmstead in northern Italy

  Pavol Micolajcak建筑事务所着眼于当地的传统，修复并扩建了南蒂罗尔的一座传统农业住宅。

  买下菲尔德霍夫农庄的年轻主人从一开始就意识到了它的独特性。整座建筑由两个统一的结构组成，根据山坡的轮廓略微偏移。它保存完好，是意大利南蒂罗尔艾萨克山谷山坡上"成对农屋"的典型代表，拥有用石头加重的木瓦屋顶的住宅和拥有令人印象深刻的陡峭茅草屋顶的棚屋，都是早期生活的真实再现。在搬进来两年后，主人决定保留房子的原始形式，并保留小棚屋原来饲养动物的功能。不过，房屋需要融入现代生活的气息，并在当代生活和历史真实性之间寻求平衡。住宅与棚屋的整体结构不能受到第三个结构的干扰。同时，要在旧的生活区和扩建结构中创建一个功能单元。

  建筑地块的地形为部分在地下的新生活区提供了机会。新"砖形"建筑直接与西向地块下面原有住宅的居住层相连，起到连接新旧建筑的作用。由此产生的结构从东向西像砖结构一样延伸，在有坡的一面几乎完全隐藏在山地景观中。只有两扇与草地齐平的天窗指明隐藏的地下空间，但建筑在朝南的一侧却表现出强有力的姿态：混凝土框架使地面打开，长长的玻璃立面使住宅朝向山谷开放。

  由于拥有宽敞、有趣的多面立面，新的起居室和卧室可以俯瞰艾萨克山谷和高耸的白云石山峰。开放的空间、超高的高度和大面积的天窗与历史建筑精致封闭的结构形成了令人兴奋的对比。阳光从天窗洒入房间，几何形状构造了令人兴奋的开放空间。此外，谷仓后面高大落叶松的美景也为建筑锦上添花。

  楼梯连接新旧建筑，连接车库层和住宅，代表了历史和现代的交汇。所有使用的材料都体现了这一概念，也均与周围的景观完美融合。天然石材、裸露混凝土、钢材和木材表面是特别为该项目选择的。由裸露混凝土建成的几何形状天花板模仿了在施工过程中移除的山顶的轮廓，而室内大量使用了木质家具和经典表面。

  菲尔德霍夫住宅扩建结构展示了如何用现代的、高质量的住宅组成部分来扩展历史建筑，同时又不影响受保护结构的原始特征。建筑师采用先进的技术，使新扩建部分和旧建筑融合在一起，它们的融合令人兴奋不已，拍手称赞。

东立面 east elevation

Pavol Micolajcak looks to vernacular traditions to restore and extend a traditional farming residence in South Tyrol.

The young owner who bought the Felderhof farmstead was conscious of its unique nature from the very beginning. The ensemble, consisting of two uniform structures, was built slightly offset in accordance with the hillside profile. Preserved in pristine condition, it is a prime example of the "pair farmstead" typical of the slopes of the Eisack Valley of South Tyrol, Italy. The residential house – with its stone-weighted wooden shingle roof, and the shed – with its impressive steep thatched roof, are both authentic representations of life in an earlier age. Two years after moving in, the owner decided to conserve the original form of the house and retain the shed for its original role of housing animals. However, some aspects of modern living are to be incorporated and a balance is sought between contemporary living and historical authenticity. The house/shed ensemble is not to be disturbed by any third structure. At the same time, a functional unit is to be built from the old living area and the extension.

The topography of the plot offers an opportunity for new living areas partly underground. The new "brick-shaped" building is designed to join directly the living floor of the existing house below the west-facing terrain, connecting old and new. The resulting structure extends in a brick

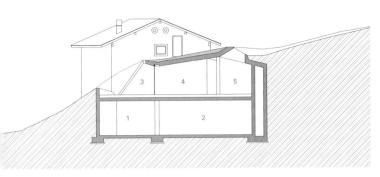
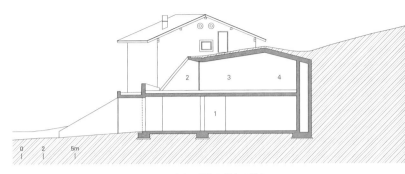

1. 车库 2. 工作室 3. 露台 4. 起居室 5. 厨房
1. garage 2. workshop 3. terrace 4. living room 5. kitchen
A-A' 剖面图 section A-A'

1. 车库 2. 露台 3. 卧室 4. 浴室
1. garage 2. terrace 3. bedroom 4. bathroom
B-B' 剖面图 section B-B'

项目名称：Felderhof House / 地点：Villanders, South Tyrol, Italy / 事务所：Pavol Mikolajcak
客户：Thomas Erlacher / 施工监理：Elisabeth Erlacher / 静态学设计：Ruben Erlacher / 电气：Elektro Oberrauch GmbH / 艺术干预：Roland Moroder
建筑面积：380m² / 设计时间：2014—2015 / 施工时间：2015~2017 / 摄影师：©Oskar Da Riz (courtesy of the architect)

shape from east to west, and on the sloped side is almost completely hidden within the mountain landscape. Only two skylights lying flush in the meadow point to the hidden underground cubature. Towards the south, however, the building offers a powerful gesture: a concrete frame opens out the ground and the long glass frontage orients the residential area towards the valley.

Thanks to the spacious, playfully faceted facade, the new living and sleeping rooms enjoy a spectacular view over the Eisack Valley and the lofty peaks of the Dolomites. Open spaces, generous heights and skylights present an exciting contrast to the delicate and closed fabric of the historic building. The skylights also allow for rooms to be flooded with sun and the geometry creates exciting and open environments. An added bonus is provided by the excellent views of the large larch trees behind the barn.

The staircase – which connects the old building with the new, and the garage floor with the residence – represents the intersection between history and modernity. All the materials used give expression to this idea and fit with the surrounding landscape. Natural stone, exposed concrete, steel and wood surfaces are the highly specialized choice made for the project. While the geometric ceiling of exposed concrete imitates the contour of the hilltop removed during construction, extensive use is made of wooden furniture and classic surfaces.

The extension of the Felderhof demonstrates how a historic building can be augmented by a modern, high-quality residential component without disturbing the original character of the protected structure. Through the techniques used, the new extension and the old building enter into an exciting, complimentary discourse.

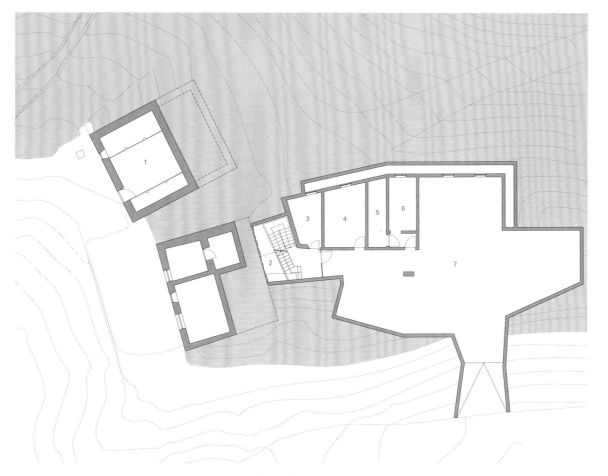

地下一层 first floor below ground

1. 马厩
2. 石墙
3. 储藏室
4. 工作室
5. 技术设备室
6. 锅炉房
7. 车库
8. 卧室
9. 书房
10. 地板
11. 厨房
12. 走廊
13. 浴室
14. 起居室
15. 衣橱
16. 露台

1. stable
2. stone wall
3. storage
4. workshop
5. technical room
6. boiler room
7. garage
8. bedroom
9. study room
10. floor board
11. kitchen
12. corridor
13. bathroom
14. living room
15. wardrobe
16. terrace

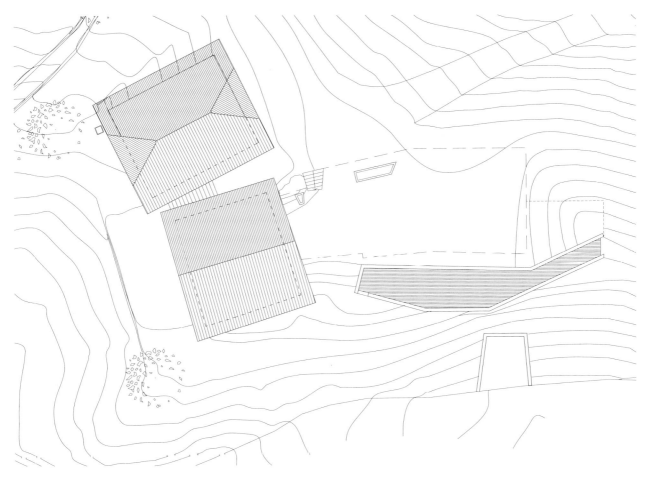

屋顶 roof

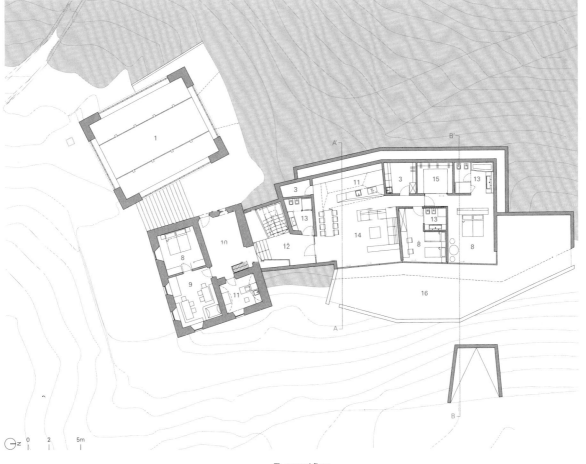

一层 ground floor

(不)熟悉的地方

(Un)famil

Finding Meaning in N

在非常规的经验中寻找意义

我们研究的这四个项目可以说是古怪的。它们给人的第一印象是独特体验的承诺,这一承诺一方面是对它们所在地方和环境所做出的探索的解释,另一方面也体现了技术力量、智能或生态敏感性的壮举。通常我们对这些应该感觉很熟悉,但是,几番品读后,你会发现每一座这样的建筑都将挑战它们所宣称的意图和意义。也许这就是为什么它们会如此令人兴奋,为什么它们会让人觉得探索它们更有挑战性。它们所具有的这种模糊性,有点像一种邀请,让参观者们用不容易获得的工具来理解他它们——就如同本文的读者。

The four projects we examine may arguably stand as oddities. What they offer, on first impression, is the promise of unique experiences; they hold this promise one time as an interpretive exploration of the place and the context they are built upon, and another as a feat of technical prowess, intellectual ability, or ecological sensibility. Normally this should feel very familiar; but, given the benefit of a second or third reading, each of these buildings will challenge their professed intentions and meanings. Perhaps this is why they feel so exciting, why they feel so challenging to explore them more. It is this ambiguity that they carry, somewhat like an invitation to decipher them with tools that aren't readily available to their visitors or – for that matter – the readers of this review.

印记_The Imprint / MVRDV
UCCA沙丘艺术博物馆_UCCA Dune Art Museum / OPEN Architecture
Tij观测站_Tij Observatory / RAU Architects + RO&AD Architecten
AZULIK Uh May艺术中心和IK实验室_AZULIK Uh May and IK LAB / Roth (Eduardo Neira)

(不)熟悉的地方——在非常规的经验中寻找意义
(Un)familiar Places – Finding Meaning in Non-Routine Experiences / Angelos Psilopulous

iar Places
-Routine Experiences

我们认为，这些建筑承载的不只是一个命题，而是一个承诺；它们让我们摆脱先入之见，让我们在建筑中重新发现我们的身体、我们的思想、我们的本性。奇怪的是，它们成功地完成了它们的挑战，不是通过实现这个承诺，而是让我们暂停，直到我们发现我们只是在状态和情况之间、在意图和影响之间、在叙述和经验之间摇摆。正是这种即将实现的感觉，那种希望得出结论的感觉，激励着我们去发现更多。

For all we argue, these buildings carry not a proposition but a promise, to throw us off our preconceptions, to have us rediscovering our bodies, our minds, our nature within their architecture. Oddly, they succeed in their challenge not by delivering on this promise but by keeping us suspended, until we find that we've merely been oscillating between states and situations, between intentions and affects, between narratives and experiences. It is this sense of being almost there, of hoping to reach a conclusion, that teases us to discover more.

(不)熟悉的地方——在非常规的经验中寻找意义
(Un)familiar Places – Finding Meaning in Non-Routine Experiences

Angelos Psilopulous

那些接触过我们当代建筑的人，现在应该已经对设计策略有了一些直观的了解，这些设计策略为我们的建筑提供了信息，就像我们为本文收集的这些项目一样。在某些情况下，这些策略主要依赖于通过一个三步设计过程对原有空间进行有条不紊的处理；有时，它们会依赖于对特定环境的重新解释，从而创造出新的内容，或者通过在熟悉的背景下并置新形式，或者通过将现有的材料和技术与新的方法和技术融合。由于建筑业定期出版物已经对这些设计进行了广泛甚至是详尽的报道，因此它们现在应该已经变得不那么重要了……

然而，不知何故，建筑总是能给我们带来新的惊喜。直到今天，像这样的建筑仍然可以嘲笑那些试图用通用术语定义它们的人，并向我们展示它们如何抵制那些试图拿分类评估来得出结果的人。有些人会很轻易地选择聚焦于建筑物的某一个特征，以证实其主张，因为这样的特征很普遍，然而，一旦我们开始拿这些建筑跟其他建筑进行对比，即使是它们在这篇文章中呈现的无实体形式，它们似乎也能展示一种不可思议的能力，吸引并邀请我们更多地探索它们。这不过是一种承诺的暗示，暗示着它们不仅仅是各部分的总和。因为这个原因，它们将激发人们这种感觉：我们需要越过它的表象，去体验和理解建筑物的意义。我们在很多方面对它们熟悉，但它们却试图承载一种陌生体验的承诺。也许这就是我们今天在判定一个成功的建筑尝试时所需要的所有标记。

乍一看，本文介绍的建筑看起来像是模仿作品：MVRDV的印记（166页）将真实地反射其周围景物；位于图卢姆罗斯由Roth（Eduardo Neira）设计的AZULIK Uh May艺术中心和IK实验室（208页）均使用天然建筑材料使游客感觉他们好像身处大地中；Tij观测站（194页）由RAU建筑师事务所和RO&AD建筑事务所设计，它将以巢中蛋的形象而存在；而OPEN建筑事务所设计的UCCA沙丘艺术博物馆（180页）则给人一种它是用海滩上的沙子堆砌起来的有趣感觉。

Those exposed to our contemporary architecture should have, by now, some intuitive familiarity with the design strategies that inform buildings such as the ones that we have collected for this article. In some cases, these strategies rely on the predominantly methodical appropriation of existing space through a "1-2-3 steps" design process; and sometimes they will rely on a reinterpretation of a given context that creates new content, either through the juxtaposition of new forms set against familiar backgrounds or through the fusion of established materials and techniques with new approaches and technologies. In the light of the extensive – even exhaustive – coverage of these designs get from industry and regular press alike, they should be getting just about trivial by now…
And yet, somehow, architecture never fails to surprise us anew. To this day, buildings like these can still tease and toy with anyone who tries to define them on generic terms, and show us how they can still show resistance against their inclusion in categorical assessments so dear to those who will attempt them. Someone could easily choose to focus specifically on one of their characteristics or the other in order to substantiate such claims, as they are indeed all of them in any place; yet once we start assimilating these buildings, even in the disembodied form that they carry on the pages of this review, they seem to reveal an uncanny ability to entice and invite us to explore them more. This is but a hint of a promise, that they are more than the sum of their parts. And for this reason, they will excite this feeling that we need to experience them and understand them for more than they are willing to reveal on first read. They are in many ways familiar, and yet they manage to carry the promise of an unfamiliar experience. Perhaps this is all the token one needs today to characterize an architectural venture as successful.
At first glance, the buildings in our feature look like an imitation: the Imprint (p.166) by MVRDV will physically reflect its immediate surroundings; the AZULIK Uh May and IK LAB (p.208) in Tulum by Roth (Eduardo Neira) will use natural materials to make visitors feel as if they inhabit the earth itself; the Tij Observatory (p.194) by RAU Architects and RO&AD Architecten will literally stand in as a nested egg, while the UCCA Dune Art Museum (p.180) by OPEN Architecture evokes the playful worldmaking afforded in the sands of its coastline.

再次解读就会发现，这些建筑要依靠现有的文脉来承载它们的意义。印记项目将其简单的体量形式操作与周围建筑的印记联系起来——从绘图板绘图开始就是这样操作的——方法是将周围建筑的线性线条雕刻在自己的建筑立面上。AZULIK Uh May艺术中心和IK实验室将通过自然来塑造它们的封闭空间。Tij观测站将用一个保护性的生态系统类型外壳来证明它外壳的有效。UCCA沙丘艺术博物馆雕刻出了一个精致复杂的内部结构，建筑师通过熟练地运用建筑基地原有沙丘的拓扑结构，将天空、海岸和大海的迷人景色引入了内部空间。

更进一步解读就会发现，这些建筑本身就是巨大的成就。Tij观测站使用参数化算法实现了402个部件的最有效的结构设计。而AZULIK Uh May艺术中心和IK实验室大胆地将物质转化为想象的意志，以提升设计师的原始艺术视野，使其具有超凡脱俗的感观和体验意识。UCCA沙丘艺术博物馆在负空间里用水泥块雕刻了一座建筑。最后，印记项目设法实现了与外部空间及其邻近建筑的开放和富有表现力的联系，尽管它本身必须满足作为一个封闭空间的要求。为了实现这一目标，建筑师发明了一种独特的设计语言，在仿制品[1]和当代艺术作品强大的独特性之间华丽地摇摆。[2]

在这里，我们意识到我们所收集的独特建筑项目是通过一个悖论联系在一起的，在这个悖论中，一个发人深省的学术话语和一个非常规的既有体验的承诺并置，从而揭示了这些建筑的意义。所有这些似乎都能让伪与真产生共鸣，让它们所产生的叙述以及它们所带来的影响产生共鸣。它们的创造者将它们的设计作为一种解释性的构造，就像一个启发式的过程，旨在通过这个过程理解建筑所处的环境，有时是真正的场地，有时是被要求实现的功能，有时是渗透在一些最特殊的建筑实例中的实验精神。但是，最终，这个设计过程也产生了一种修辞，创造了新的意义，但更好的是新的内容，我们的思想、身体和感官都在这个修辞中得到了满足。优秀建筑的

Second reading shows these buildings to rely on existing context in order to carry their meanings. The Imprint will drape its simple volumetric formal manipulations with the imprint of the buildings around it – a literal crossing over from the drawing board – by physically carving their linear tracing onto its facade. The AZULIK Uh May and IK LAB will harvest nature to shape their enclosed space. The Tij Observatory will justify its shell with the terms of a protective ecosystem type enclosure. And the UCCA Dune Art Museum will carve an intricate complex of interiors, combined with their mesmerizing visual reliefs towards the sky, the coast and the sea, by navigating skillfully the topology of the existing dunes of the building site.

Third reading shows that these buildings stand as achievements in their own right. The Tij Observatory uses parametric algorithms to achieve the most efficient structuring of its 402 parts assembly. On the other hand, the AZULIK Uh May and IK LAB will boldly bend matter to the will of the imagination in order to elevate their designer's original artistic vision to an otherworldly sensorial and experiential awareness; and the UCCA Dune Art Museum will carve a building, with its cement masses, out of negative space. Finally, The Imprint manages to achieve an open and expressive connection to the outside space and its neighboring buildings albeit having to oblige to the requirement of being an enclosed space in itself. For this to happen, the architects invent a unique design language that oscillates magnificently between pastiche[1] and the powerful singularity of a contemporary work of art.[2] This is where we realize that our collection of unique buildings is connected through a paradox, in that one discovers their meanings through the juxtaposition of a thought-provoking intellectual discourse and the promise of a non-routine existential experience. All of them seem to resonate the pseudo and the genuine alike, the narrative that they generate and the affect that they bring. Their creators use their design as an interpretive construction, much like a heuristic process by which they aim to understand the context they are called to build upon; sometimes it's the genius loci of the place, sometimes it's the program they are asked to carry out, and sometimes it is just this spirit of experimentation that imbues some of the most special instances of architecture. But, in the end,

作品就是这样的，不是吗？但是，我们收集的项目似乎与它们的作者的自主关注产生了共鸣，也与它们周围领域的外部性产生了共鸣。它们将通过将现象学经验的预期规则与一种智慧的技巧融合在一起，方便地为参观者提供了一种身处建筑室外的体验。最后，在我们的手中是一组模糊的对象，它们在以一种简单的方式顽固地抵制我们的解读：在印记项目令人眼花缭乱的视觉效果中，我们的思维能够穿越现实，同样，在参观AZULIK Uh May艺术中心和IK实验室以及UCCA沙丘艺术博物馆的洞穴和浮雕空间时，我们的身体也能够超越感官漩涡；最后，Tij观测站可以被同时从两个方面进行解读，一个是它的象征性质，另一个是它作为一个巢穴的真实存在。

Timotheus Vermeulen和Robin van den Akker将这种模糊性解释为当代文化生产的一个基本特征，在"现代热情和后现代讽刺"[3]之间摇摆不定。实际上，我们在这里研究的所有项目都为给定的问题提出了良好的解决方案，然而，它们中没有任何一个会真正宣称，它们已经信心十足地解决了这个问题。反之亦然，所有项目都会很容易地提出一种解释，甚至是一种反讽，来呈现它们最初的设计纲要，然而它们所提供的叙述最终都充满了一种积极的精神，就好像它们确实旨在改善我们对享受设计空间的既定成见。也许是大量的环境危机、金融危机和社会危机促使建筑师和设计师在他们的方法中采取肯定的立场，或者更确切地说，是情感上的立场，但是他们的建筑仍然对他们意图的严肃性产生了质疑。正如卢克·特纳所言："我们认为这表现为一种明智的天真，一种实用主义的理想主义，一种温和的狂热主义，在真诚与讽刺、解构与建构、冷漠与情感之间摇摆，试图获得某种超越的地位，仿佛这样的事情尽在我们的掌握之中。"[4]

对于那些不愿深入研究文化理论和批评的人来说，它看起来只像是一个陌生的领域在进行着熟悉的主题：坚持当代建筑重新诠

this design process also produces a rhetoric that creates new meanings – better yet, new content – upon which our minds, our bodies and our senses all feast. This is how good architecture works after all, isn't it? However so, our projects seem to resonate the autonomous concerns of their authors as well as the externalities of their surrounding fields on equal doses, and they will handily offer their visitors an out-of-body experience by fusing the expected set of rules of a phenomenological experience with an intellectual intrigue. In the end, what we have in our hands is a set of ambiguous objects that stubbornly resist to be read in one simple way: one's mind is equally able to transcend reality in the dizzying visuals of The Imprint as one's body would transcend physicality in the sensuous vortexes of the AZULIK Uh May and IK LAB and the transposing caves and reliefs of the UCCA Dune Art Museum; and finally, the Tij Observatory can be interpreted equally for its symbolic nature and its physical presence as a nest.

Timotheus Vermeulen & Robin van den Akker explain this ambiguity as a fundamental characteristic of contemporary cultural production, oscillating between "a modern enthusiasm and a postmodern irony".[3] Indeed, all of the projects we examine here propose a well-defined solution to a given problem; yet none of them will actually claim that they solve this problem with the confidence of a universal truth. And vice-versa: all of them will handily propose an interpretive, even an ironic, take on their original brief, yet the narrative they offer is ultimately imbued in a positive spirit, as if they do indeed aim to improve our established preconceptions about the enjoyment of designed space. Perhaps it is the abundant environmental, financial, and societal crises that drive architects and designers nowadays to take affirmative – or, rather, affective – positions in their approaches, but their architecture still resonates a doubt about the seriousness of their intentions. As Luke Turner puts it: "We see this manifest as a kind of informed naivety, a pragmatic idealism, a moderate fanaticism, oscillating between sincerity and irony, deconstruction and construction, apathy and affect, attempting to attain some sort of transcendent position, as if such a thing were within our grasp."[4]

To those who will not delve deep into cultural theory and criticism, it simply looks like an unfamiliar take on a familiar theme: the persistence of contemporary architecture to reinterpret, perhaps even reinvent, the commonplace. And for those of us who get to savor the experience these oscillating, ambiguous objects promise, we may

释甚至是彻底改造平凡的东西。而对于我们中那些享受摇摆不定、模棱两可的事物所带来的体验的人来说，很可能正经历一场毫无疑问的魔法之旅。事实上，我们敢说，这些地方的设计师和游客都是被想要尝试这种体验的简单愿望所驱使。引用尤金·奥尼尔《更庄严的大厦》中一个被滥用的段落：

> ……我们被一个童话故事迷住了，花了一生去寻找一扇魔法门和一个失落的和平王国，我们被一个贪婪的骗子从那里赶了出来。[5]

但就像剧中的角色一样，掌握这个王国决定性钥匙的既不是建筑师，也不是客户，更不是用户和不相关的路人。相反，他们是联系在一起的，是这个异想天开的炼金术游戏的共谋者，在意图和机会之间、行动和反应之间、设计和消费之间摇摆，而且这些是同时进行的，直到渴望被熄灭，如果这样的事情会发生的话。

我们在本部分中研究的所有项目都要求与原有情况形成和谐的关系，但它们有足够的探索能力把这种关系转变成某种独特的发人深省的东西。在每个项目中，对看似特定的环境——形式、空间和地点、文化和历史、自我和人——的感知意味着从一个特定的有利点重新构建。只有这个有利点在根据建筑各相关方的偶然动态不断变化，有时是无意的。我们认为正是这件事激发了这些项目的活力：这种模棱两可的感觉产生于人与这些空间之间的互动，产生于建筑物的预期性质和它们所能产生的梦想之间，这是一种几乎就在那里的感觉。

very well be in for an unapologetically magical ride. Indeed, we would venture that both the designers and the visitors of these places are driven by the simple hope of such an experience. To quote one rather abused passage from Eugene O'Neill's *More Stately Mansions*:

(…) obsessed by a fairy tale, we spend our lives searching for a magic door and a lost kingdom of peace from which we have been dispossessed by a greedy swindler. [5]

But same as for the characters in the play, it looks like neither architect nor client, not even the users or the innocent passers-by hold a definitive key to this kingdom. Instead they remain entangled, co-conspirators to this whimsical game of alchemy, swaggering between intention and occasion, acting and reacting, designing and consuming, all at once, all until the yearning is extinguished, if ever such a thing may come to be.
All of the projects we examine in our feature claim to form a harmonious relationship with the existing situation; and yet, they are explorative enough to turn this relationship into something uniquely thought-provoking. The perception of the seemingly particular context in each project – of form, of space and place, of culture and history, of self and people – is meant to be reconstituted from a specific vantage point. Only this vantage point keeps shifting, sometimes unintentionally, according to the haphazard dynamics of this mix of co-conspirators. We would argue that this very thing is what energizes these projects: this sense of ambiguity generated by the interaction between people and these spaces, between the intended nature of the buildings and the dreams that they can generate, this sense of being almost there.

1. Either by echoing the adjacent buildings of the Paradise City project on their façade or by nudgingly hinting towards James Wines's BEST supermarkets in the 80's by "lifting" the building's outer shell to form an entrance.
2. We would argue that the "singular objects" of Baudrillard and Nouvel still seem to inform the building's foundational origins. *The Singular Objects of Architecture* (Minneapolis: University of Minnesota Press, 2002); but then one cannot escape a reference to Venturi and Scott-Brown's distinction between "ducks" and "sheds", i.e. buildings that embody their function and buildings that their skin forms an autonomous narrative. *Learning from Las Vegas: The Forgotten Symbolism of Architectural Form*, 17th print (Cambridge, Mass.: The MIT Press, 2000).
3. Timotheus Vermeulen and Robin van den Akker, "Notes on Metamodernism," *Journal of Aesthetics & Culture 2*, no. 1 (January 2010): 5677, https://doi.org/10.3402/jac.v2i0.5677.
4. Luke Turner, "Metamodernism: A Brief Introduction," *Notes on Metamodernism* (blog), December 12, 2015, http://www.metamodernism.com/2015/01/12/metamodernism-a-brief-introduction/
5. Eugene O'Neill, *More Stately Mansions: The Unexpurgated Edition* (Oxford University Press, 1988), p.281.

印记
The Imprint

MVRDV

MVRDV 设计的戏剧性和反射性的印记项目迎接飞来首尔的旅客
MVRDV's theatrical and reflective 'Imprint' greets travelers flying into Seoul

MVRDV完成了天堂城印记项目的建设，这是一个新的由两座建筑组成的艺术和娱乐综合体，靠近首尔的仁川机场。一座建筑(Chroma)中有一个夜总会，另一座建筑(Wonder box)中有一个室内主题公园，这两个无窗的结构中有三个关键的设计元素：周围建筑特色立面的印记，抬高的入口，以及覆盖夜总会建筑一角的黄金入口点。

MVRDV设计的印记项目是大型天堂之城综合项目的一部分，该综合项目由6栋建筑组成，提供全套娱乐设施和酒店服务，距离韩国最大的机场不到1km。对于本案两座建筑的规划，客户要求设计没有窗户，但仍与综合项目内其他建筑相融合的结构。印记项目的设计源于一个简单的问题：一个富有表现力的立面，即使没有窗户，也能与周围的环境产生联系吗？该设计通过投影周围建筑的外立面来实现这一点，周围建筑的外立面像影子一样覆盖在新建筑简单的建筑形式和广场上，像浮雕图案一样印在新建筑的外立面上。

MVRDV的负责人和联合创始人Winy Maas说："通过将周围的建筑原样布置在我们建筑的立面和中心广场上，我们使印记项目与周围环境建立了联系。这可以确保一致性。天堂之城不是像拉斯维加斯那样的独立建筑的集合，而是一座真正的城市。"

为了实现周围建筑的理想"印记"，立面采用玻璃纤维增强混凝土板建成。由于3869块面板中有许多都是独一无二的，所以在设计阶段需要使用MVRDV的3D建模文件单独制作模具。安装完成后，这些板材被漆成白色，以凸显浮雕的效果。

Winy Maas解释说："两个月前，大部分的覆层已经完成，客户说，'这是一件艺术品'。有趣的是，他们正在寻找这种动力——娱乐可以变成艺术，或者建筑可以变成艺术。那么，建筑与艺术的区别是什么呢？我认为抽象是这个项目的一部分，但它必须让人惊讶，吸引人，再让人平静下来。"

黄金点是该项目最明显、最引人注目的表达元素，甚至吸引了即将在仁川机场降落的乘客的目光。金色的实现很简单，采用的是金色的涂料，夜晚，立面上的灯光亮起，更增加了金色的效果。大多数立面都是从底部照明的，而黄金点则是从顶部照亮的，凸显了这一侧立面。

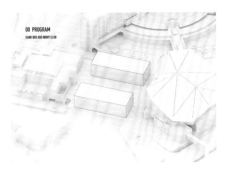

项目名称：The Imprint / 地点：Incheon, Korea / 事务所：MVRDV
负责人：Winy Maas / 合伙人：Wenchian Shi / 设计团队：María López Calleja, Daehee Suk, Xiaoting Chen, Kyosuk Lee, Guang Ruey Tan, Stavros Gargaretas, Mafalda Rangel, Dong Min Lee
合作建筑师：Gansam Architects＆Partners / 立面顾问：VS-A Group Ltd
镶板顾问：Withworks / 玻璃纤维增强水泥：Techwall / 照明：L'Observatoire International
客户：Paradise Segasammy Co., Ltd. / 用途：cultural, bar, restaurant / 用地面积：9,800m²
建筑面积：club–2,421.73m²; wonder box–3,095.90m² / 总楼面积：club–15,122.05m²; wonder box–10,343.58m² / 结构：reinforced concrete, steel frame, steel encased reinforced concrete
材料：GFRC, painting / 施工时间：2016—2018
建筑师：©Ossip van Duivenbode (courtesy of the architect)

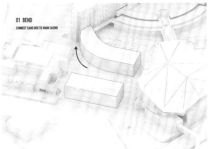

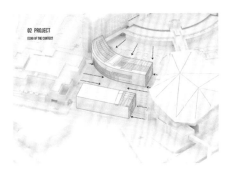

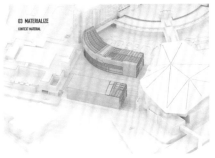

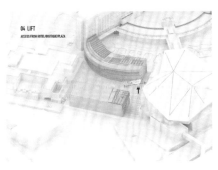

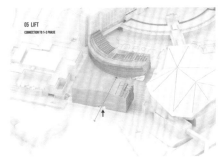

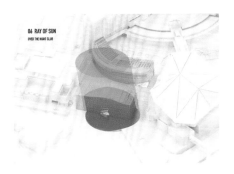

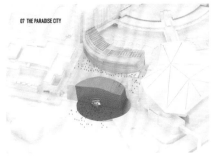

"这座新建筑获得了一抹金色,好像入口在晚上也会被一缕阳光照亮,"Maas说,"即将降落的飞机上的乘客可以从海上看到这个'太阳',它仿佛在表达欢迎来到韩国。"

入口处的立面像窗帘一样升起,露出镜面天花板和玻璃媒体地板,散发出一种由内而外的兴奋感。

Mass总结道:"因此,反射和戏剧性是相结合的。在我们的设计中,在夜间的恶作剧之后,白天会有一种禅意般的寂静,为聚会后的人们提供一种几乎真实的反思环境。我想,乔治·德·基里科会非常喜欢画它的。"

MVRDV has completed construction on The Imprint at Paradise City, a new two-building art and entertainment complex in close proximity to Seoul's Incheon Airport. Featuring a nightclub in one building (Chroma) and an indoor theme park in the other (Wonderbox), the windowless structures feature three key design elements: imprints of the facade features of surrounding buildings, lifted entrances, and a golden entrance spot covering one corner of the nightclub building.

MVRDV's The Imprint is part of the larger Paradise City complex of six buildings providing a full suite of entertainment and hotel attractions less than a kilometer away from Korea's largest airport. Given the proposed program of the two buildings, the client required a design with no windows, yet one that still integrated with the other buildings in the complex. The design of The Imprint therefore arises from a simple question: can an expressive facade connect with its surroundings even though it has no windows?

The design achieves this by projecting the facades of the surrounding buildings, which are "draped" over the simple building forms and plazas like a shadow, and "imprinted" as a relief pattern onto the facades.

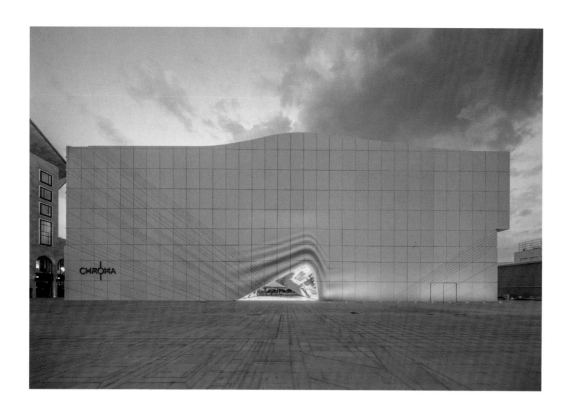

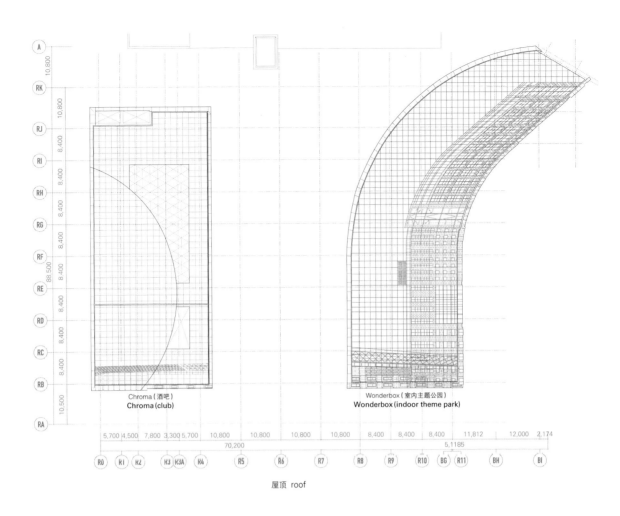

屋顶 roof

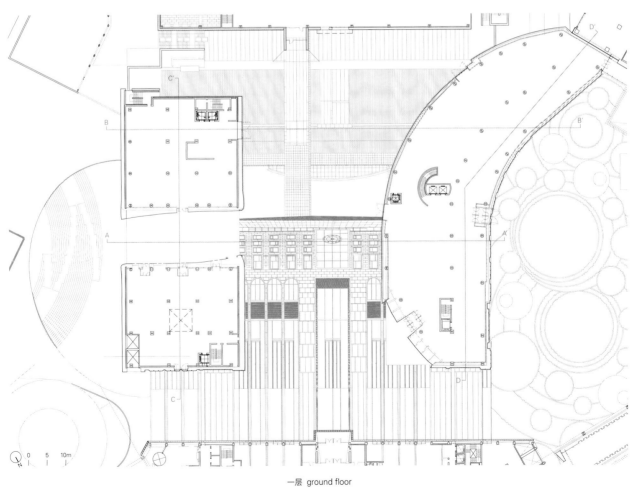

一层 ground floor

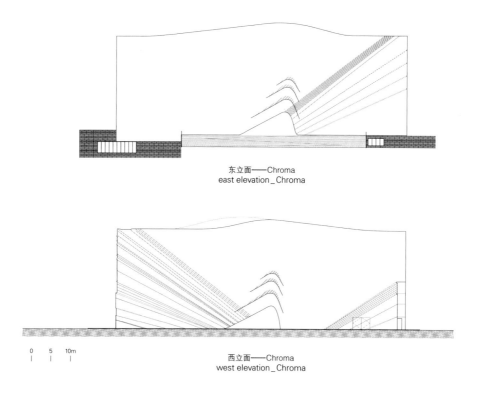

东立面——Chroma
east elevation_Chroma

西立面——Chroma
west elevation_Chroma

0 5 10m

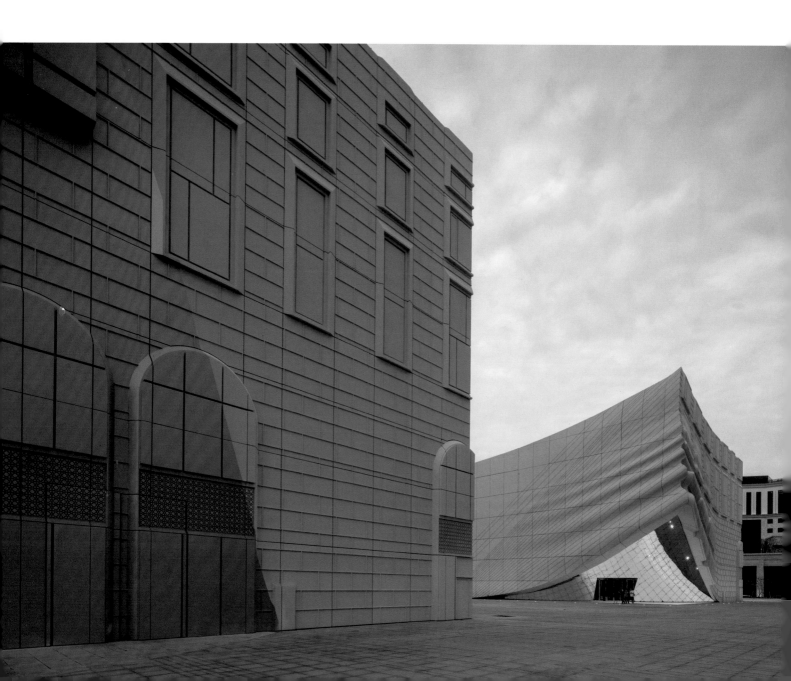

东立面——Wonderbox
east elevation_Wonderbox

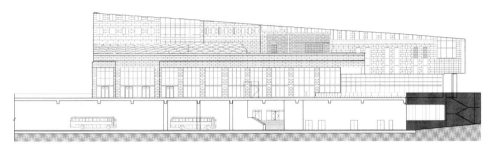

西立面——Wonderbox
west elevation_Wonderbox

"By placing, as it were, surrounding buildings into the facades of our buildings and in the central plaza, we connect The Imprint with the neighbors," says Winy Maas, principal and co-founder of MVRDV. "This ensures coherence. Paradise City is not a collection of individual objects such as Las Vegas, but a real city."

In order to achieve the desired "imprint" of the surrounding buildings, the facade is constructed of glass-fiber reinforced concrete panels. As many of the 3,869 panels are unique, the construction required molds to be individually produced using MVRDV's 3D modelling files from the design phase. Once installed, these panels are painted white in order to emphasize the relief.

As Winy Maas explains: "Two months ago most of the cladding was done and client said, 'this is an art piece'. What is interesting is that they are looking for that momentum – that entertainment can become art or that the building can become artistic in that way. What, then, is the difference between architecture and art? The project plays with that and I think that abstraction is part of it, but it has to surprise,

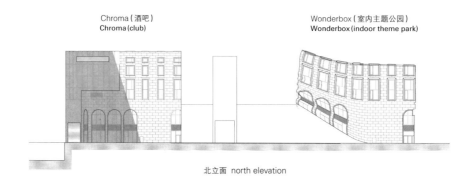

北立面 north elevation

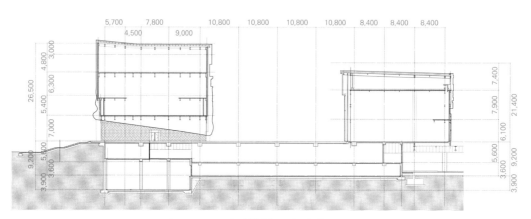

A-A' 剖面图 section A-A'

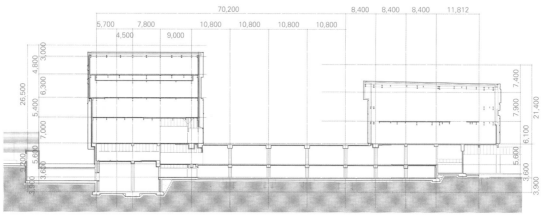

B-B' 剖面图 section B-B'

seduce and it has to calm down."

The golden spot is the project's most obvious and attention-grabbing expressive element, even catching the eyes of passengers coming into land at Incheon Airport. The golden color is achieved simply, in gold paint, and is reinforced by the lighting of the facades at night: while the majority is lit from below, the gold spot is highlighted from above.

"The virgin building has received a splash of gold. This makes it as if the entrance is also illuminated at night by a ray of sunlight," says Maas. "Passengers in the incoming aircraft can already see this "sun" from above the ocean, as a kind of welcome to Korea."

The entrances, where the facades are lifted like a curtain to reveal mirrored ceilings and glass media floors, exude a sense of the excitement happening inside.

"Reflection and theatricality are therefore combined," concludes Maas. "With our design, after the nightly escapades, a zen-like silence follows during the day, providing an almost literally reflective situation for the after parties. Giorgio de Chirico would have liked to paint it, I think."

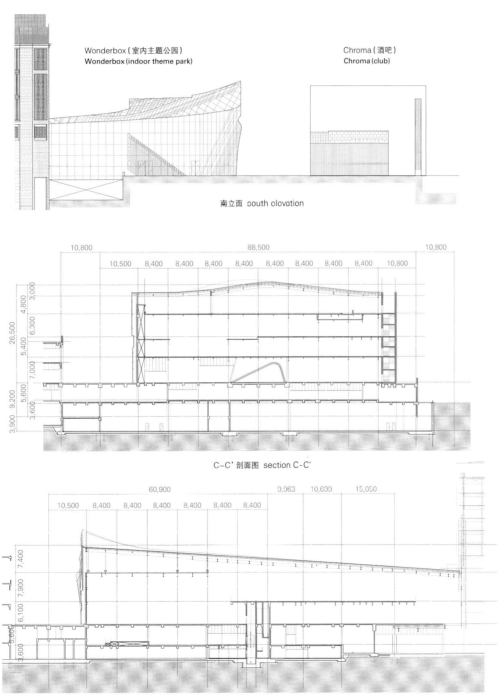

UCCA 沙丘艺术博物馆
UCCA Dune Art Museum

OPEN Architecture

UCCA 沙丘艺术博物馆——原始空间被雕刻在沙丘下面
UCCA Dune Art Museum – primeval spaces are carved beneath the dune

在中国北部渤海湾沿岸的一个安静的海滩上，有一座新建的UCCA沙丘艺术博物馆，它被雕刻在沙子里，并悄悄地消失在那里。

常年的风把海滩上的沙子吹成了沿岸几米高的沙丘，沙丘经由低矮的灌木和其他地面覆盖物固定住了。博物馆就建在这个沙丘下面。被沙子包裹的空间相互连接，有着有机的形状，类似于洞穴——人类的原始家园，洞穴墙壁曾经是人类最早的艺术作品的画布。隐藏在大海和沙滩之间的沙丘艺术博物馆设计简单、纯粹、动人，是对原始和永恒的空间形式的回归。

在沙丘下创建艺术博物馆的决定源于建筑师对自然深深的崇敬和保护脆弱的沙丘生态系统的愿望，这些脆弱的沙丘生态系统是由数千年的自然力量形成的。由于博物馆的存在，这些沙丘将被保留下来，而不是像沿岸许多其他沙丘那样被夷为平地，为海景房地产项目开发腾出空间。

一系列细胞状的连续空间容纳了大小不一的画廊、一间咖啡馆和一些附属空间。在穿过一条长而暗的隧道和一个小小的接待区后，空间突然打开，游客就进入了最大的多功能画廊。在那里，一束来自上方天窗的阳光安静而强烈地充满了整个空间。透过建筑的不同开口，博物馆游客可以在一天中看到天空和大海不断变化的表情。一部螺旋楼梯通向沙丘顶部的瞭望台，引导好奇的观众从下面的黑暗角落来到广阔的开放空间。脚下，博物馆就像一个与身体和灵魂亲密接触的隐蔽的避难所，一个深思自然和艺术的地方。

沙丘艺术博物馆的混凝土外壳复杂的三维几何形状是由秦皇岛当地的工人（其中一些人以前是造船工人）手工塑造的，使用的是由小木条制成的模板，有时在需要更紧密的模型时还会使用一些更有弹性的材料。建筑师有意保留了模板留下的不规则和不完美的纹理，让人们能够感受到和看到建筑手工建造的痕迹。此外，建筑的门窗、接待台、吧台和浴室水池都是现场定制设计和制作的。咖啡馆里的八张桌子也是由建筑师设计的，每一张都有独特的形状，与八个主要画廊空间的平面布局相匹配。

建筑有许多天窗，每一个都有不同的方向和大小，全年为博物馆空间提供经过精心调节的自然光。覆盖沙子的屋顶大大降低了建筑的夏季热负荷。低能耗、零排放的地源热泵系统取代了传统空调。

On a quiet beach along the coast of Bohai Bay in the northern part of China, the UCCA Dune Art Museum is carved into the sand, where it gently disappears.

Countless years of wind have pushed the beach's sand into a dune along the shore, several meters high, and stabilized by low-rising shrubs and other ground cover. The museum lies beneath this dune. Enveloped by sand, its interconnected, organically-shaped spaces resemble caves – the primeval home of man, whose walls were once a canvas for some of humanity's earliest works of art. Hidden between the sea and the sand, the design of the Dune Art Museum is simple, pure, and touching – a return to primal and timeless forms of space.

The decision to create the art museum underneath the dunes surrounding it was born out of both the architects' deep reverence for nature and their desire to protect the vulnerable dune ecosystem, formed by natural forces over thousands of years. Because of the museum, these sand dunes will be preserved instead of leveled to make space for ocean-view real estate developments, as has happened to many other dunes along the shore.

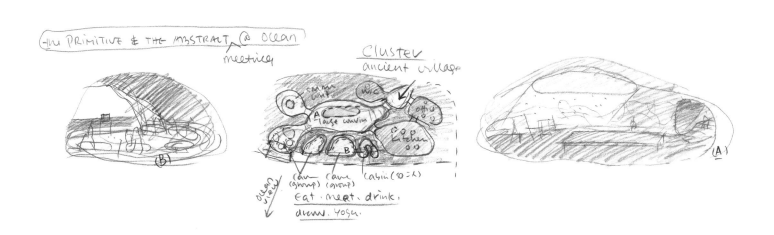

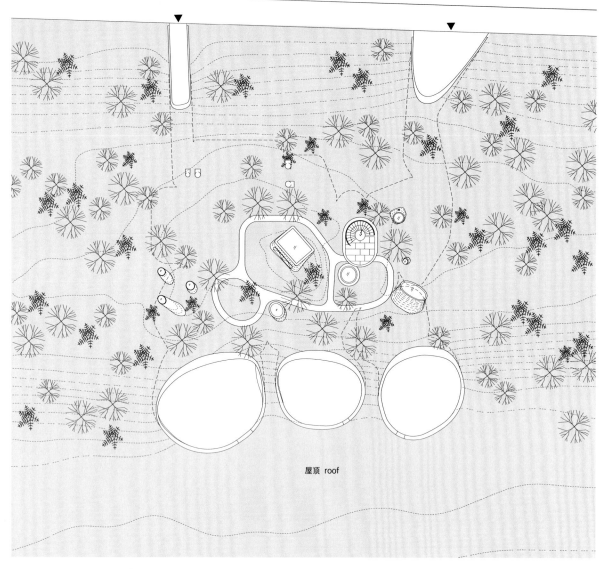

屋顶 roof

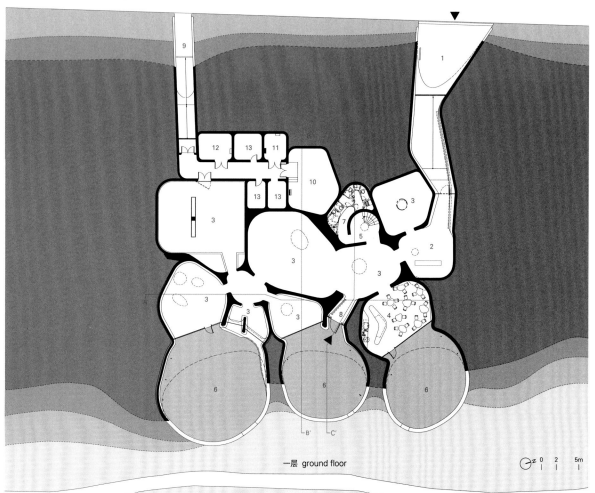

一层 ground floor

1. 主入口
2. 大厅
3. 画廊
4. 咖啡馆
5. 台阶
6. 室外展区
7. 卫生间
8. 次入口
9. 设备入口
10. 地热电站
11. 弱电间
12. 电气室
13. 备用室

1. main entrance
2. lobby
3. gallery
4. cafe
5. stairs
6. outdoor exhibition
7. toilet
8. secondary entrance
9. service entrance
10. geothermal plant
11. low voltage room
12. electrical
13. spare room

186

画廊空间天窗
gallery space skylight

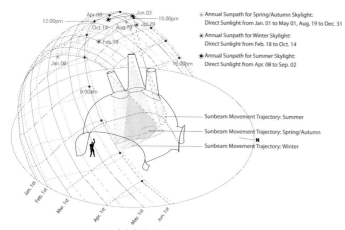

全年直射阳光
direct sunlight throughout the year

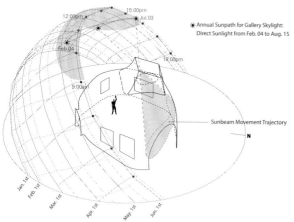

4月18日至8月15日的太阳直射，展示墙上没有直射阳光
direct sunlight from April 18th to August 15th
no direct sunlight on the display wall

咖啡馆天窗
cafe skylight

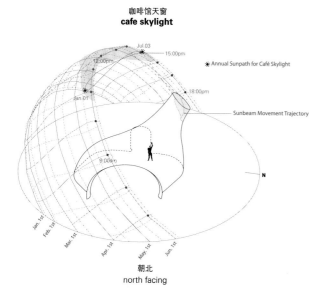

朝北
north facing

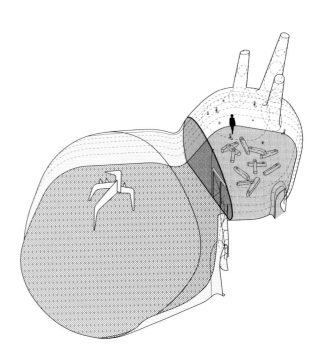

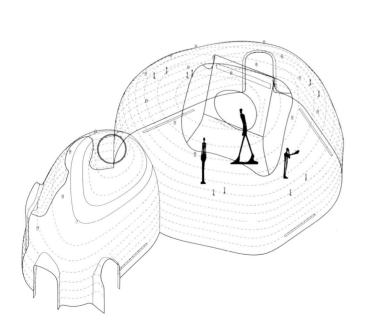
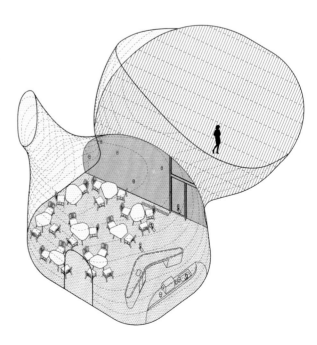

项目名称：UCCA Dune Art Museum / 地点：Qinhuangdao, China / 事务所：OPEN Architecture / 主要负责人：Li Hu, Huang Wenjing / 项目团队：Zhou Tingting (Project architect), Wang Mengmeng, Hu Boji, Fang Kuanyin, Joshua Parker, Lu Di, Lin Bihong, Ye Qing, Steven Shi, Jia Han / 当地设计机构：CABR Technology Co., Ltd. / 照明设计：X Studio, School of Architecture, Tsinghua University, China, OPEN Architecture / 运营：UCCA / 客户：Aranya / 功能：reception, cafe, community room, exhibition spaces, outdoor exhibition, roof terrace / 建筑面积：930m² / 设计时间：2015 /竣工时间：2018
摄影师：©Tian Fangfang (courtesy of the architect) - p.181, p.185, p.190, p.192; ©WU Qingshan (courtesy of the architect) - p.183, p.184, p.188~189, p.191, p.193

A series of cell-like contiguous spaces accommodate differently-sized galleries, a cafe, and some ancillary spaces. After passing through a long, dark tunnel and a small reception area, the space suddenly opens up as visitors enter the largest multifunctional gallery. There, a beam of daylight from the skylight above silently yet powerfully fills the space. Looking through different openings framed by the building, museum-goers can observe the ever-changing expressions of the sky and sea throughout the day. A spiral staircase leads to a lookout on top of the sand dune, guiding curious audiences from the dark recesses below to the vast openness above. Underfoot, the museum emerges as a hidden shelter, intimate to the body and soul – a place to thoughtfully contemplate both nature and art.

The complex three-dimensional geometry of the Dune Art Museum's concrete shell is shaped by hand by local workers in Qinhuangdao (some of whom are former shipbuilders), using formwork made from small strips of wood, and occasionally some more elastic materials when tighter curvatures are needed. The architects deliberately retain the irregular and imperfect texture left by the formwork, allowing traces of the building's manual construction to be felt and seen. In addition, the building's doors and windows, reception desk, bar counter, and bathroom sinks are all custom-designed and made on site. The eight tables in the cafe are also designed by the architects, each with a distinct shape matching that of the floor plans of the eight main gallery spaces.

The building's many skylights, each with a different orientation and size, provide carefully tempered natural lighting for the museum's spaces at all times of the year; its sand-covered roof greatly reduces the building's summer heat load; and a low-energy, zero-emission ground source heat pump system replaces traditional air conditioning.

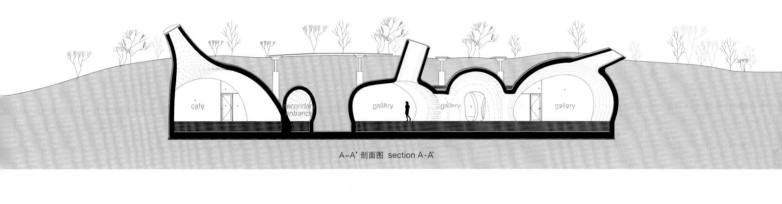

A-A' 剖面图 section A-A'

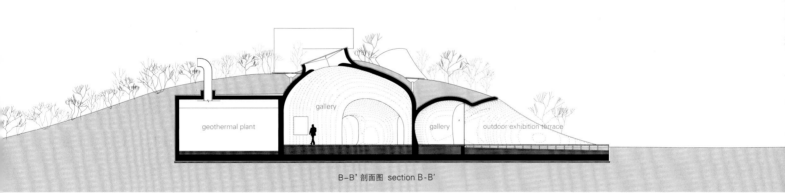

B-B' 剖面图 section B-B'

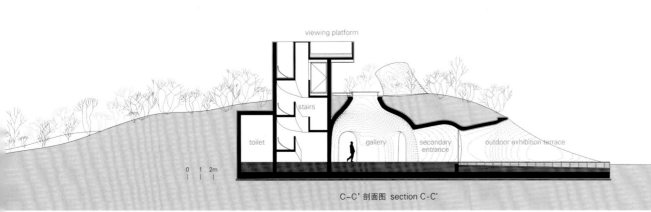

C-C' 剖面图 section C-C'

Tij 观测站
Tij Observatory

RAU Architects + RO&AD Architecten

蛋形观测站，观察鸟类生活的可持续木结构
An egg-shaped observatory; sustainable wooden construction for observing bird-life

从这个鸟类观测站，游客可以看到荷兰哈林弗里特三角洲独特的自然风光和丰富的鸟类生活。除了其他生物，还有成千上万的大型燕鸥在舍尔胡克海岸外的小岛上筑巢，新设立的鸟类观测站可以俯瞰这些小岛。

由于使用了可持续和可循环的材料，且结构形状独特，这个完全可重建的巨型木结构为自然爱好者提供了一个真正体验野生动物生活的场地。

这个雕塑般的木结构被命名为"Tij"，意思是"潮汐"，当然，还有荷兰语蛋形设计的意思。

大自然的艺术品

观测站由芦苇、栗树木杆和沙子组成，其参数化的设计使结构和观察孔的形状、构造和大小保持了良好的关系。舍尔胡克自然保护区的当地芦苇被作为覆盖层用在了建筑上。大跨度的木材使用的是文件到工厂的Zollinger施工方法。之前使用过的隔板被用来建造通往观测站的隧道。

Tij的首席建筑师Thomas Rau说："由于Tij拥有可完全重建的能力，而且采用模块化和实体化设计，因此完全满足可持续结构的所有关键要求，具有循环使用的潜力。我们建造建筑的原则就是所有东西都可以拆解而不会失去其价值，从而确保了对生态系统造成的压力是最小的。观测站的形状非常特别，模仿了大型燕鸥的蛋。是自然本身产生了这种形状。"

自然爱好者可以从Tij观测站通过木材混凝土复合楼板充分观赏繁殖岛、哈林维利特丹和周围地区。蛋的底部由固雅木木梁构成。在水位比较高的时候，观测站的底部会被水淹没。水面以上的部分是用松木建造的。

未来

随着鸟类观测站的开放，参与该项目的自然组织也完成了Droomfonds项目，但他们将继续在哈灵水道共同努力，虽然Droomfonds时期已经结束。六个参与合作的自然组织看到了进一步恢复哈林弗列特湾及其周围动态生机的机会，并将愉快地在近年来该区域内已经共建的良好基础上，继续进行建设。

From this bird observatory, visitors have a fantastic view of the exceptional nature of the Haringvliet Delta, in the Netherlands, and its rich bird life. Among other things, a colony of thousands of large terns nest on the small islands off the coast of Scheelhoek, which is overlooked by the new bird observatory.

Thanks to the use of sustainable and circular materials, and the structure's unique shape, the fully rebuildable huge wooden construction offers nature-lovers a genuine experience of wildlife.

The sculptural wooden entity was given the name "Tij", which means "tide" and, of course, the egg-shaped design, in Dutch.

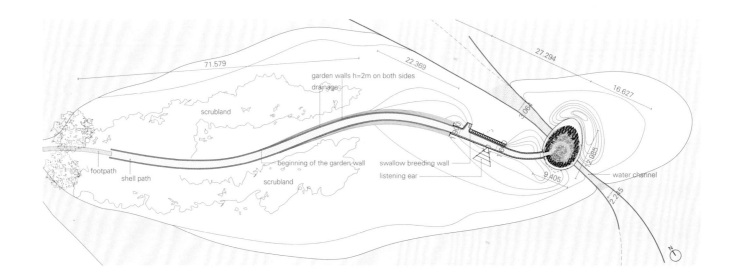

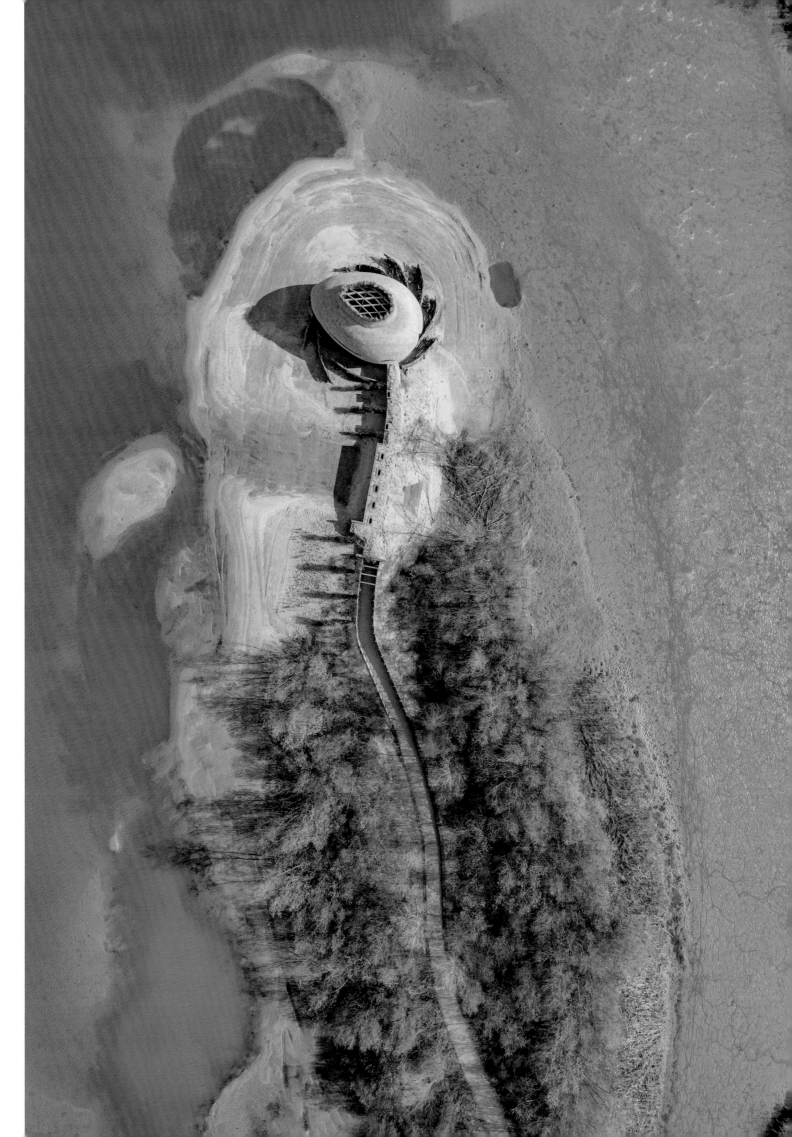

北立面 north elevation

Nature artwork

The observatory, made of reeds, chestnut poles and sand, is parametrically designed to achieve a good relationship between shape, construction and size, of the structure and the viewing-holes. Local reeds from the Scheelhoek nature reserve have been used for the covering. Large spans of wood are made using a file-to-factory Zollinger construction method. Pre-used bulkheads are used to make a tunnel to the observatory.

Thomas Rau, chief architect of Tij said: "Thanks to its complete rebuilding capabilities, modularity and materialization, Tij fully meets all the key points for a sustainable structure with circular potential. By building everything in such a way that everything can be taken apart without losing any of its value, we ensure that the strain on the ecosystem is minimal. The shape of the observatory is extra special, mimicking the egg of the large tern. Nature itself produced this shape."

From Tij, nature lovers can enjoy an impressive view of the breeding islands, the Haringvlietdam and the surrounding area via a hybrid wood-concrete floor. The bottom part of the egg is made from Accoya wooden beams. This section of the observatory can be submerged under water in cases where the discharge level of the river is high. The part that remains above water is constructed from pine.

Future

With the opening of the bird observatory, the participating nature organizations complete the Droomfonds project, but they will continue to join forces in the Haringvliet, even after the Droomfonds period has ended. The six cooperating nature organizations see many opportunities for further restoring the dynamics in and around the Haringvliet inlet, and are happy to continue building on the sound basis that they have been able to lay down together within the region in recent years.

南立面 south elevation

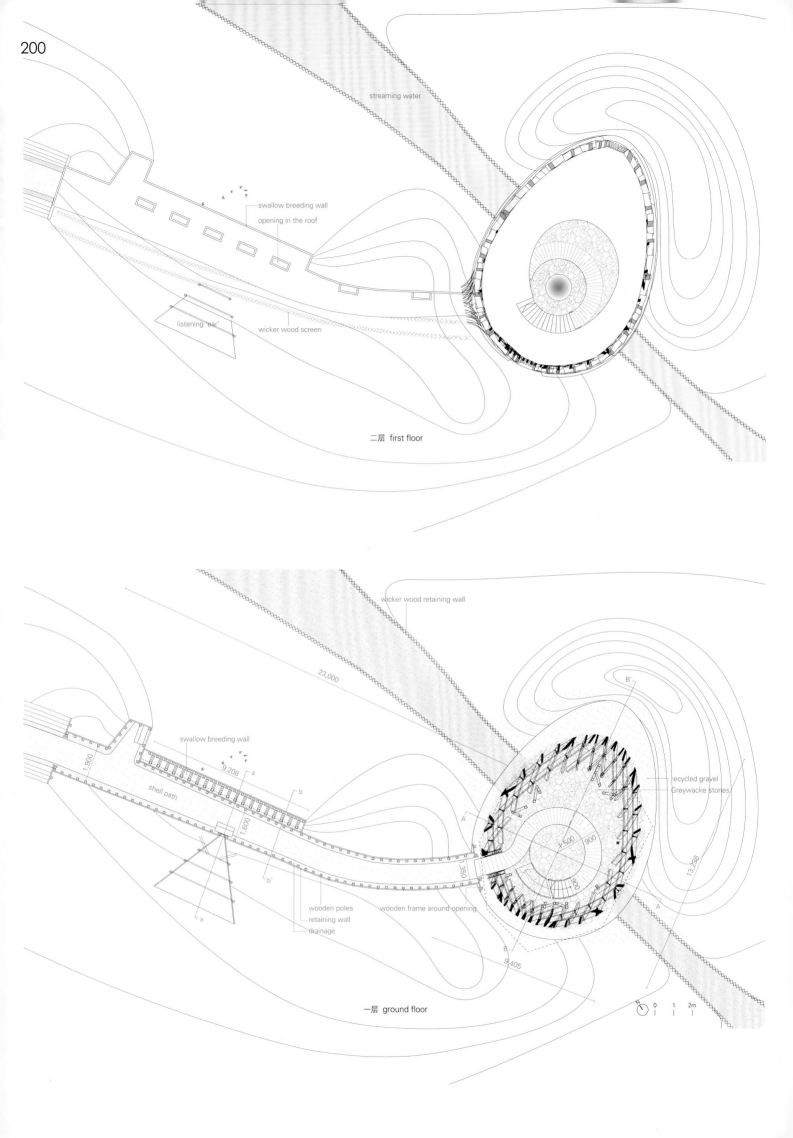

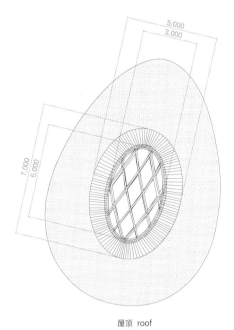

屋顶 roof

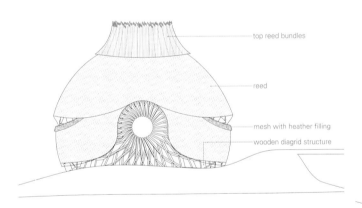

北立面 north elevation

- top reed bundles
- reed
- mesh with heather filling
- wooden diagrid structure

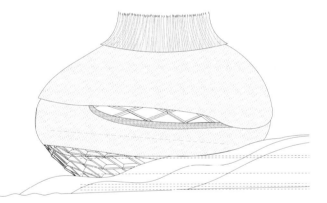

西立面 west elevation

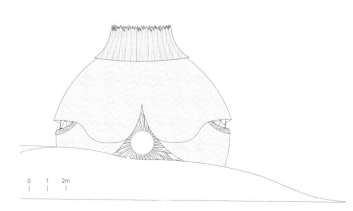

南立面 south elevation

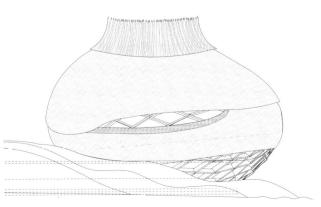

东立面 east elevation

项目名称：TIJ Observatory / 地点：Scheelhoek, Stellendam, the Netherlands / 事务所：RAU in collaboration with RO&AD Architects
项目团队：Thomas Rau, Ad Kil, Ro Koster, Michel Tombal, Jochem Alferink, Martin van Overveld, Athina / 施工建议：General and foundation_Breedid, the Hague
木结构：Aalto University, Finland / 木材工程公司：Geometria, Finland / 承包商：Van Hese Infra, Middelburg
委托方：Bird protection Netherlands/Natuurmonumenten, Zeist / 客户：Bird Protection Netherlands

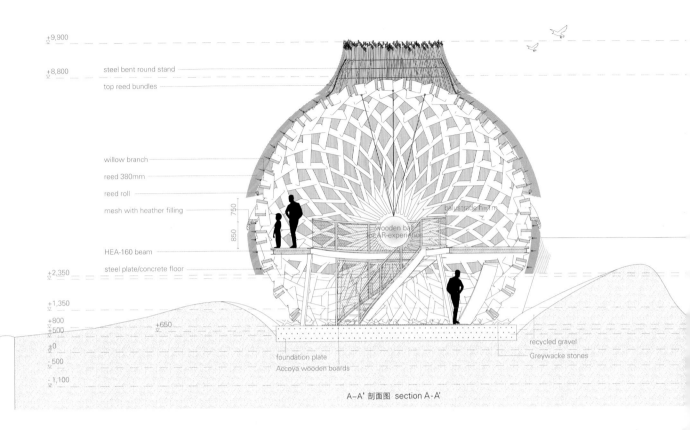

A-A' 剖面图 section A-A'

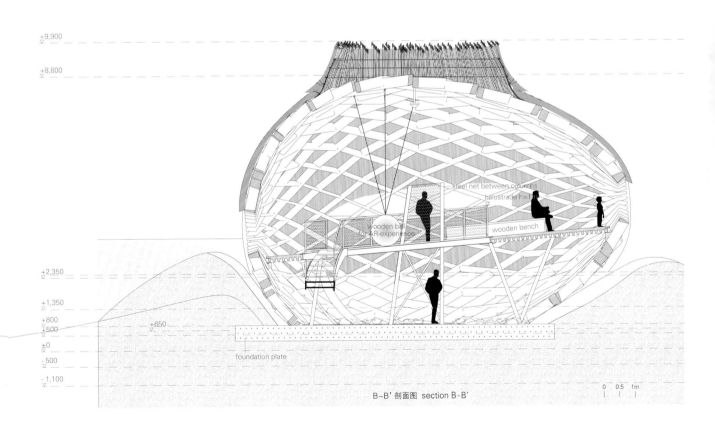

B-B' 剖面图 section B-B'

用途：Bird observatory / 用地面积：5,000 m² / 建筑规模：two stories above ground
结构：concrete floor on wooden columns and wooden Zollinger system in facade / 室外饰面：Reed / 室内饰面：Wood, concrete, steel / 材料：wood, concrete, steel / 造价：EUR 500,000~1,000,000 / 设计时间：2016.9—2018.7 / 施工时间：2018.8—2019.2
摄影师：©Katja Effting (courtesy of the architect) - p.198~199, p.202, p.206~207 right; courtesy of the RO&AD Architecten - p.195, p.196~197

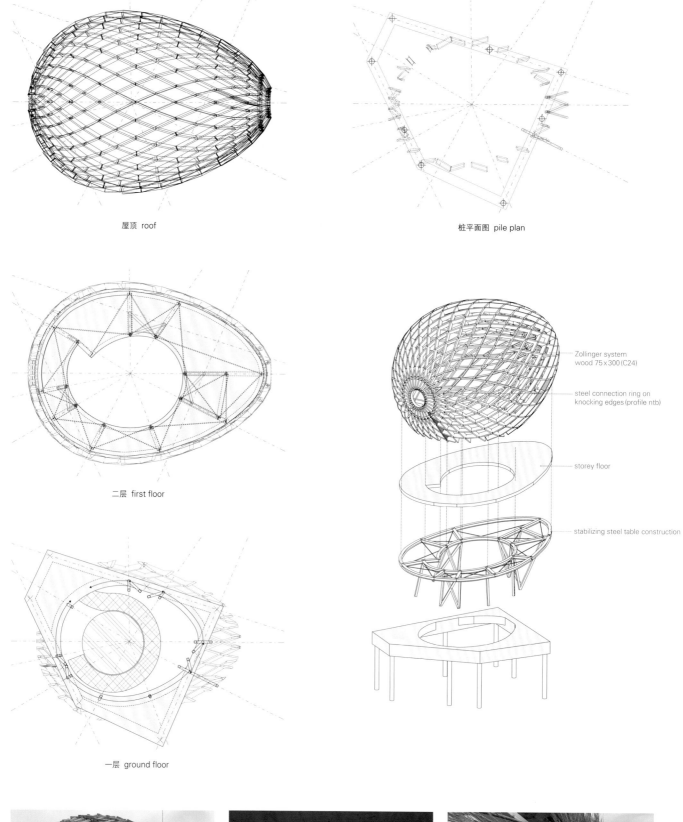

屋顶 roof

桩平面图 pile plan

二层 first floor

一层 ground floor

Zollinger system
wood 75 × 300 (C24)

steel connection ring on
knocking edges (profile ntb)

storey floor

stabilizing steel table construction

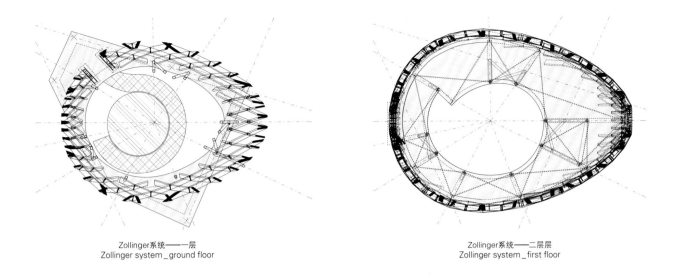

Zollinger系统——一层
Zollinger system _ ground floor

Zollinger系统——二层层
Zollinger system _ first floor

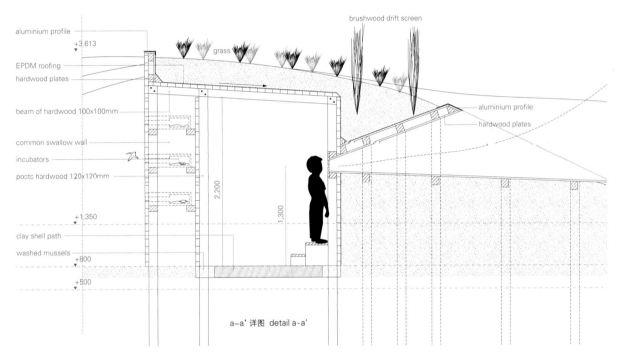

a-a' 详图 detail a-a'

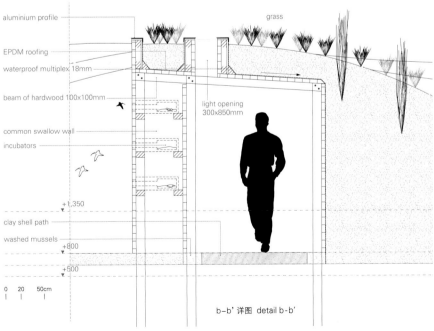

b-b' 详图 detail b-b'

AZULIK Uh May and IK LAB
Roth (Eduardo Neira)

生态灵感，捕捉尤卡坦半岛丛林的灵魂
Ecologically inspired, capturing the soul of the Yucatan Peninsula jungle

AZULIK Uh May艺术中心是一个多立面的灵活综合体，位于尤卡坦半岛丛林的中心，设计灵感来自丰富的自然环境和当地丰富的玛雅文化遗产。

AZULIK Uh May艺术中心包含一系列创意空间，包括一个创新艺术空间、一个专注于时尚和设计的实验室、一所艺术和工艺学校、一个先进的录音棚，以及为来自不同领域的艺术家和创新思想领袖们提供住宿的住所。

AZULIK Uh May艺术中心由社会企业家Roth (Eduardo Neira) 创立，在他的生态灵感度假胜地图卢姆AZULIK和邻近的艺术空间图卢姆IK实验室的愿景上进行了拓展："AZULIK Uh May艺术中心将会鼓励和支持我们遇到这个时代最好的精神——致力于探索个人和社区的新型连接，统一于土著部落与环境和谐相处的共同愿望，帮助他们在有众多表现形式的艺术的庇护下迎接当代生活的挑战。"

AZULIK Uh May试图通过对材料和技术的精心选择来展示对保护和整合环境的承诺，拟人的结构是建筑的特点，这些结构看起来就像是自然地从地面上升起，通过浮桥和蜿蜒的小路彼此连接。第一个结构中是一个16m高的由混凝土和白粉藤皮制成的圆顶，圆顶上有生命之花——由重叠的圆圈组成的花状图案几何形状——而它的基础是根据斐波那契数列的比例呈现的。与它相邻的是Roth自己的新家，为相遇和聚会提供了空间。

Roth说："AZULIK Uh May艺术中心抓住了这个地方的灵魂，经过精心设计，可以保护和拥抱当地的生态系统。建造过程中建筑师考虑到了碳足迹，没有砍伐一棵树来为建筑腾出空间，而是让这些结构将原有的植被包围起来。"

AZULIK Uh May的中心是一所专注于艺术和工艺通用语言的学校，它将当地的玛雅人、常驻艺术家、国际学生和学者聚集在一起。通过创造力的经验分享，学校将培养和创立出一种社区归属感。

AZULIK Uh May艺术中心还将包括结合西医与千年玛雅治疗实践的医疗设施，以及探索、保存和弘扬玛雅祖先烹饪智慧的研究厨房和餐厅。AZULIK Uh May将继续开发项目来支持其现有设施，使项目不断发展扩大，它将包括一个艺术博物馆 (由容纳场地特定设施的展馆组成)、一个再循环和可持续研究中心以及一个新技术中心。

A-A' 立面 elevation A-A'

0 2 5m

Inspired by the abundant natural environment and the rich heritage of the local Mayan culture, AZULIK Uh May is a multifaceted and flexible complex, in the heart of the jungle of the Yucatan Peninsula.
AZULIK Uh May encompasses an array of creative spaces, including an innovative art space, a lab dedicated to fashion and design, an arts and crafts school, a state of the art recording studio as well as residencies for artists and innovative thought leaders from a diverse range of fields.
Founded by social entrepreneur Roth (Eduardo Neira), AZULIK Uh May expands upon the vision reflected in his ecologically inspired resort AZULIK Tulum and the adjacent art space IK LAB Tulum: "AZULIK Uh May will encourage and favor the encounter of the finest spirits of our times committed to exploring new forms of reconnecting as individuals and as a community alike, united by the shared desire to learn from the native tribes to live in harmony with the environment and to help them navigate the challenges of contemporary life under the aegis of art in a myriad of manifestations."

项目名称：AZULIK Uh May and IK LAB / 地点：Francisco Uh May, Q.R., Mexico
建筑师：Roth (Eduardo Neira) / 合作者：AZULIK, IK LAB, Enchanting Transformation
用途：visual arts center / 项目时间：2018.12 / 摄影师：courtesy of the Enchanting Transformation

AZULIK Uh May attempts to demonstrate a commitment to preserving and integrating the environment through carefully considered choices of both materials and techniques. The anthropomorphous structures, characteristic of his architecture, appear to naturally arise from the ground and are connected with each other through floating bridges and meandering paths. Among the first of the structures is a 16m-high concrete and bejuco dome with a cupola crowned by the Flower of Life – a geometrical shape composed of overlapping circles arranged in a flower like pattern – while its base unfolds according to the proportions of the Fibonacci sequence. Adjacent to this lies Roth's own new home, which provides a space for encounters and gatherings.

"AZULIK Uh May captures the soul of the location and has been carefully designed to preserve and embrace the local ecosystems. The construction process is mindful of the carbon footprint and not a single tree has been cut to make

space for the buildings, instead the structures embrace the existing vegetation," says Roth.

At the heart of AZULIK Uh May, a school focusing on the universal language of art and craft will bring together the local Mayan population, artists in residence, international students and scholars. Through a shared experience of creativity, the school will foster and develop the sense of belonging in a community.

AZULIK Uh May will moreover encompass medical facilities that integrate Western medicine with millennia-old Mayan healing practices, as well as a research kitchen and restaurant that will explore, preserve and honor the healing virtues of ancestral Mayan culinary wisdom.

AZULIK Uh May will continue to develop a program to support its current facilities, evolving and expanding to include an art museum composed of pavilions housing site-specific installations, a recycling and sustainability research center and a hub for new technologies.

P62 OMA
Rem Koolhaas founded OMA in 1975 together with Elia and Zoe Zenghelis and Madelon Vriesendorp. Graduated from the Architectural Association in London, and is a professor at Harvard University. He co-heads the work of both OMA and AMO, the research branch of OMA, operating in areas beyond the realm of architecture. Current projects include the Taipei Performing Arts Centre, a new building for Axel Springer in Berlin, and the Factory in Manchester. Ellen van Loon joined OMA in 1998 and has been a partner since 2002. Has led several award-winning building projects that combine sophisticated design with precise execution. Some of her most significant contributions include the new G-Star Headquarters in Amsterdam (2014) and the Dutch Embassy in Berlin (2003), winner of the European Union Mies van der Rohe award in 2005. Iyad Alsaka joined OMA as a director in 2007 and became partner in 2011 and is responsible for OMA's work in the Middle East and Africa. Born in 1969 in Syria, Iyad holds a degree in Architectural Engineering from the University of Aleppo.

P110 Marià Castelló
Graduated from the School of Architecture of Barcelona in 2002 with Honors. His life and his work are closely linked to the island of Formentera, whose landscape, culture and tradition have been an important source of inspiration. The fruit of his work has been recognized at European level by being a Finalist of the XIV Spanish Architecture Biennial (2018), winner of NAN Spanish National Award (2018), the Best Architects 20 Award (2019), and the Awards of Architecture of Ibiza and Formentera (2018). Numerous publications and exhibitions have spread his work internationally.

P90 ELEMENTAL
Is a group of architects, founded in 2001 and led by Alejandro Aravena and partners Gonzalo Arteaga, Juan Cerda, Víctor Oddó and Diego Torres, based in Santiago de Chile. Is a 'Do Tank' company, the strength of which is the innovation and design quality of public interest and social impact; working in urban projects of infrastructure, public space, transportation and housing, operating within the city and its capacity to generate wealth and quality of life, using it as a shortcut towards equality.
Alejandro Aravena, CEO of Elemental, is a member of the Pritzker Prize Jury since 2009. He also was a 2016 Pritzker Prize Laureate and named Honorary RIBA International Fellow in 2009. He is director of 15th Venice Biennale International Architecture Exhibition 2016.

P16 Valerio Olgiati
Was born in 1958, Chur and studied architecture at ETH Zurich. In 1996 he opened his own practice in Zurich and later in 2008 together with his wife Tamara in Flims, Switzerland. The first time he received attention was in 1999 with the museum The Yellow House in Flims. His most important buildings include the Schoolhaus in Paspels and visitor center for the Swiss National Park in Zernez. The major solo exhibition of his work took place in 2012 at MoMa Tokyo. Has taught at ETH Zurich, the Architectural Association, and at Cornell University, as well as held the Kenzo Tange Chair at Harvard University. Since 2002 he has been a full professor at the Academy of architecture, Mendrisio and University of Lugano.

P30 Foster + Partners
Is an international studio for architecture, engineering and design, led by Founder and Chairman Norman Foster and a Partnership Board. Founded in 1967, the practice is characterized by its integrated approach to design, bringing together the depth of resources required to take on some of the most complex projects in the world. Over the past five decades the practice has pioneered a sustainable approach to architecture and ecology through a strikingly wide range of work, from urban masterplans, public infrastructure, airports, civic and cultural buildings, offices and workplaces to private houses and product design. The studio has established an international reputation with buildings such as the world's largest airport terminal at Beijing, Swiss Re's London Headquarters, Hearst Headquarters in New York, Millau Viaduct in France, the German Parliament in the Reichstag, Berlin, The Great Court at London's British Museum, Headquarters' for HSBC in Hong Kong and London, and Commerzbank Headquarters in Frankfurt.

P84 Nelson Mota
Graduated at the University of Coimbra, and teaches architectural design and theory at Delft University of Technology. In his doctoral dissertation 'An Archaeology of the Ordinary' (Delft University of Technology, 2014) he examined the relationship between housing design and the reproduction of vernacular social and spatial practices. Is production editor and member of the editorial board of *Footprint* and a founding partner of Comoco architects.

P194 RAU Architects
Was founded by German architect, Thomas Rau(1960) in 1992, Amsterdam. Has been operating from a strong sense of awareness with respect to designing environmentally friendly buildings. Is devoted to the long-term interests of the Earth and its inhabitants aiming to minimise the CO_2 and material impact of buildings. Is one of the first studios that start focusing on circularity by designing buildings as 'raw material depots'. After graduated from the University of Bonn, Thomas Rau studied Architecture at the Alanus University of Arts and Social Sciences and RWTH Aachen University. He also founded One Planet Architecture institute (OPAi) in 2008 and Turntoo, a service company in 2010. In 2013, he was named Dutch architect of the year and received the ARC13 Oeuvre Award by designing the first circular building. Together with co-author Sabine Oberhuber, he published the *Material Matters* (2016), in which they talk about their philosophy in detail.

P195 RO&AD Architecten
Was founded in 2002 by Ro Koster (1963) and Ad Kil (1965). They work on a great variety of commissions and buildings where sustainability and making ecological positive buildings become main goals of the office. Began to gain international recognition with the Moses Bridge in the Netherlands. Has a staff of 8 people and work on a variety of projects, from big area development to very small acupuncture projects in society and landscape.

Angelos Psilopulous

P160 Angelos Psilopulous
Is architect and educator, currently working as a Senior Lecturer at the Department of Interior Architecture, University of West Attica, in Athens, Greece. Has presented and published in a variety of internationally accredited peer reviewed publications and conferences, and he has taken part as a researcher in a number of Research Programs with topics varying from the study of traditional vernacular architecture to the study of architectural competitions. He's also been working as a freelance architect since 1998, undertaking projects both on his own and in collaboration with reputed firms and architectural practices in Greece. Since 2014 he is a contributing writer and a member of the Editorial Board at C3 Magazine, Korea.

P40 **Ateliers Jean Nouvel**
Jean Nouvel was born in Fumel, France in 1945. Started his first architecture practice in 1970 and Obtained his degree at ENSBA(Ecole Nationale Supérieure des Beaux-Arts), Paris in 1972. In 1989, The Arab World Institute in Paris was awarded the Aga-Khan Prize. In 2000, he received the Lion d'Or of the Venice Biennale. In 2001, he received the Royal Gold Medal of the RIBA, the Praemium Imperial of Japan's Fine Arts Association and the Borromini Prize given to architects under 40. Was appointed Docteur Honoris Causa of the Royal College of Art in London en 2002. Also received the International Highrise Award 2006 and prestigious Pritzker Prize in 2008. In France, he received many prizes including the Gold Medal of the French Academy of Architecture, two Équerre d'Argent and National Grand Prize for Architecture.

P130 **Atelier Branco Arquitetura**
Was founded in 2012 by Matteo Arnone[left] and Pep Pons[right]. Is an architecture atelier based in Sao Paulo that develops projects of different scales and programs located around the world. Has been awarded Golden Medal of Premio T Young Claudio De Albertis by the Triennale di Milano with the project Casa Biblioteca in 2018. Matteo Arnone and Pep Pons are graduated in Master of Science in Architecture from the Accademia di Architettura Mendrisio at Università della Svizzera Italiana (2011). Matteo was a visiting professor at Istitute of European Design, Milan. Pep is grauated postgraduation lato-sensu in Housing and City by the Escola da Cidade, São Paulo (2012).

P146 **Pavol Mikolajcak Architekt**
Pavol Mikolajcak was born in 1981 in Levoca, Slovakia. Graduated in architecture from Vienna University of Technology with the thesis 'Museum of Contemporary History in Bratislava' in 2007. Worked at the studio of Christoph Mayr Fingerle in Bolzano from 2003 to 2009, focusing on architectural planning and competitions. Since 2012, has been a member of Chamber of Architects of Bolzano. Is working as freelance architect in Bolzano.

P169 MVRDV

Was set up in Rotterdam, the Netherlands, in 1993 by Winy Maas, Jacob van Rijs, and Nathalie de Vries. Engages globally in providing solutions to contemporary architectural and urban issues. The work of MVRDV is exhibited and published worldwide and receives international awards. Together with Delft University of Technology, MVRDV runs "The Why Factory", an independent think tank and research institute providing argument for architecture and urbanism by envisioning the city of the future.

P180 OPEN Architecture

Was co-founded by Li Hu[right] and Huang Wenjing[left] in New York city in 2003 and established the studio's Beijing office in 2008. Is a passionate team of designers collaborating across different disciplines to practice urban design, landscape design, architectural design, and interior design, as well as the research and production of design strategies in the context of new challenges.
Li Hu is a former partner of Steven Holl Architects. Received his B. Arch. from Tsinghua University in 1996 and his M. Arch. from Rice University in 1998. Is a visiting professor at the Tsinghua University School of Architecture and director of Columbia University GSAPP's Studio-X Beijing. Huang Wenjing is a visiting professor at the Tsinghua University School of Architecture. Received her B. Arch. from Tsinghua University in 1996, and M. Arch. from Princeton University in 1999. Both has been the recipient of '50 under 50: Innovators of the 21st Century', *ICON*'s 'Future 50', *GQ*'s 'Architect of the Year 2014'.

P4 Hans Ibelings

Is editor and publisher of *The Architecture Observer*. Lectures at the John H. Daniels Faculty of Architecture, Landscape and Design of the University of Toronto. Prior to moving to Canada in 2012, he was the editor of *A10 new European architecture*, a magazine he founded in 2004 together with graphic designer Arjan Groot. Is the author of a number of books, including *Supermodernism: Architecture in the Age of Globalization* and *European Architecture Since 1890*.

P8 Diego Terna

Graduated from Polytechnic University of Milan in 2004 and worked in the offices of Stefano Boeri and Italo Rota. In 2012 he founded the Quinzii Terna Architecture practice focusing on architecture, urbanism and research, supported by publishing, criticism and didactics. Contributes to national and international architecture magazines and websites. Taught in several universities in Italy, Central and South-America and China.

P208 Roth (Eduardo Neira)

Is the CEO and Founder of AZULIK and the President of the Foundation Enchanting Transformation. All of his initiatives aim to reconnect individuals and tribes with themselves, with one another and with the environment, involving projects in the fields of hospitality, architecture, sustainability, wellness, art, fashion, creativity and innovation, etc. With every project, Roth encourages the preservation of local culture and the recovery of knowledge and values from ancestral wisdom to promote different forms of interaction with and protection of the natural environment. Through Enchanting Transformation, he is currently developing a network of art and health schools in the Yucatan peninsula, and a 131 hectares' project in

© 2020 大连理工大学出版社

版权所有·侵权必究

图书在版编目(CIP)数据

建筑的着陆艺术 / 荷兰MVRDV建筑设计事务所等编；李璐译. — 大连：大连理工大学出版社，2020.9
ISBN 978-7-5685-2656-2

Ⅰ. ①建… Ⅱ. ①荷… ②李… Ⅲ. ①建筑艺术—世界 Ⅳ. ①TU-861

中国版本图书馆CIP数据核字(2020)第155956号

出版发行：大连理工大学出版社
　　　　　（地址：大连市软件园路80号　邮编：116023）
印　　刷：上海锦良印刷厂有限公司
幅面尺寸：225mm×300mm
印　　张：14
出版时间：2020年9月第1版
印刷时间：2020年9月第1次印刷
出 版 人：金英伟
统　　筹：房　磊
责任编辑：杨　丹
封面设计：王志峰
责任校对：张昕焱
书　　号：978-7-5685-2656-2
定　　价：298.00元

发　行：0411-84708842
传　真：0411-84701466
E-mail：12282980@qq.com
URL：http://dutp.dlut.edu.cn

本书如有印装质量问题，请与我社发行部联系更换。

墙体设计
ISBN: 978-7-5611-6353-5
定价: 150.00元

新公共空间与私人住宅
ISBN: 978-7-5611-6354-2
定价: 150.00元

住宅设计
ISBN: 978-7-5611-6352-8
定价: 150.00元

文化与公共建筑
ISBN: 978-7-5611-6746-5
定价: 160.00元

城市扩建的四种手法
ISBN: 978-7-5611-6776-2
定价: 180.00元

复杂性与装饰风格的回归
ISBN: 978-7-5611-6828-8
定价: 180.00元

内在丰富性建筑
ISBN: 978-7-5611-7444-9
定价: 228.00元

建筑谱系传承
ISBN: 978-7-5611-7461-6
定价: 228.00元

伴绿而生的建筑
ISBN: 978-7-5611-7548-4
定价: 228.00元

微工作·微空间
ISBN: 978-7-5611-8255-0
定价: 228.00元

居住的流变
ISBN: 978-7-5611-8328-1
定价: 228.00元

本土现代化
ISBN: 978-7-5611-8380-9
定价: 228.00元

都市与社区
ISBN: 978-7-5611-9365-5
定价: 228.00元

木建筑再生
ISBN: 978-7-5611-9366-2
定价: 228.00元

休闲小筑
ISBN: 978-7-5611-9452-2
定价: 228.00元

景观与建筑
ISBN: 978-7-5611-9884-1
定价: 228.00元

地域文脉与大学建筑
ISBN: 978-7-5611-9885-8
定价: 228.00元

办公室景观
ISBN: 978-7-5685-0134-7
定价: 228.00元